Springer Complexity

Springer Complexity is an interdisciplinary program publishing the best research and academic-level teaching on both fundamental and applied aspects of complex systems—cutting across all traditional disciplines of the natural and life sciences, engineering, economics, medicine, neuroscience, social and computer science.

Complex Systems are systems that comprise many interacting parts with the ability to generate a new quality of macroscopic collective behavior the manifestations of which are the spontaneous formation of distinctive temporal, spatial or functional structures. Models of such systems can be successfully mapped onto quite diverse "real-life" situations like the climate, the coherent emission of light from lasers, chemical reaction-diffusion systems, biological cellular networks, the dynamics of stock markets and of the internet, earthquake statistics and prediction, freeway traffic, the human brain, or the formation of opinions in social systems, to name just some of the popular applications.

Although their scope and methodologies overlap somewhat, one can distinguish the following main concepts and tools: self-organization, nonlinear dynamics, synergetics, turbulence, dynamical systems, catastrophes, instabilities, stochastic processes, chaos, graphs and networks, cellular automata, adaptive systems, genetic algorithms and computational intelligence.

The three major book publication platforms of the Springer Complexity program are the monograph series "Understanding Complex Systems" focusing on the various applications of complexity, the "Springer Series in Synergetics", which is devoted to the quantitative theoretical and methodological foundations, and the "Springer Briefs in Complexity" which are concise and topical working reports, case studies, surveys, essays and lecture notes of relevance to the field. In addition to the books in these two core series, the program also incorporates individual titles ranging from textbooks to major reference works.

Editorial and Programme Advisory Board

Valery I. Klyatskin

Fundamentals of Stochastic Nature Sciences

 Springer

Valery I. Klyatskin
A.M. Obukhov Institute of Atmospheric
   Physics
Russian Academy of Sciences (RAS)
Moscow
Russia

Translation by A. Vinogradov (А. Виноградов) from the Russian language edition: В.И. Кляцкин, Основы стохастического естествознания. © URSS (Lenand), Moscow, Russia. All Rights Reserved.

ISSN 1860-0832   ISSN 1860-0840 (electronic)
Understanding Complex Systems
ISBN 978-3-319-86036-7   ISBN 978-3-319-56922-2 (eBook)
DOI 10.1007/978-3-319-56922-2

Printed on acid-free paper

This Springer imprint is published by Springer Nature
The registered company is Springer International Publishing AG
The registered company address is: Gewerbestrasse 11, 6330 Cham, Switzerland

Preface

Stochastic structure formation in random media is considered using examples of elementary dynamic systems related to the two-dimensional geophysical fluid dynamics (Gaussian random fields) and to stochastically excited dynamic systems described by partial differential equations (lognormal random fields). In the latter case, spatial structures (clusters) may be formed with a probability one in almost every system realization due to rare events happening with vanishing probability. Problems involving stochastic parametric excitation occur in fluid dynamics, magnetohydrodynamics, plasma physics, astrophysics, and radiophysics. A more complicated stochastic problem dealing with anomalous structures on the sea surface (rogue waves) is also considered, where the random Gaussian generation of sea surface roughness is accompanied by parametric excitation.

Moscow, Russia   Valery I. Klyatskin

Contents

Introduction

Mankind exists in a certain realization of stochastic spatio-temporal chaos. Yet physicists believe that the basic dynamic equations of mechanics, fluid dynamics, magnetohydrodynamics, electrodynamics, acoustics, optics, radiophysics, etc. describe the actual evolution of the world in space and time.

In 2014, the scientific community celebrated a jubilee—the 450th anniversary of the birth of the great scientist Galileo [1], who contended that Nature formulates its laws in the language of mathematics. And the equations of dynamics in space and time indubitably embody one of the main manifestations of mathematics in physics! If Galileo was right, then a question arises as to how the laws of Nature can 'be rectified' from these equations without analyzing possible physical mechanisms of concrete phenomena. This question can only be answered with a rigorous statistical analysis. Namely such an approach is demonstrated in this work.

We begin by formulating the *main task* of a statistical analysis of stochastic dynamic systems in the way we understand it: *to find out, based on a relevant statistical analysis, such general properties of stochastic dynamic systems that are manifested with a unit probability, i.e., for almost all realizations of the systems being considered.* This is related to the fact that we commonly do not possess an ensemble for averaging, and specialists in numerical modeling and experimentalists alike deal only with separate realizations of random processes and fields. Traditional statistical averaging gives, as a rule, 'the mean over a hospital ward'. There are, of course, exceptions (see Sect. 1.1).

In stochastic dynamic systems described by partial differential equations in a stochastic structure formation in space and time may take place in some events with a unit probability for individual realizations of the fields involved. Such processes and phenomena, occurring with a unit probability, we will call *coherent*. This kind of 'statistical coherence' can be considered some organization of a complex dynamic system, while singling out its *statistically stable characteristics* is anal-ogous to the introduction of the concept of *coherence* understood as *self-organization* in multicomponent systems arising from chaotic interactions among their elements (see, for example, book [3]).

This work deals with three types of the simplest dynamic systems: systems related to Gaussian random fields, to stochastic parametric excitation, and to stochastic parametric excitation fed by Gaussian pumping (the combined event). They are all described by partial differential equations.

Note that even in Gaussian random fields one may encounter nontrivial situations, atypical for ordinary Gaussian noise. Such phenomena occur, for example, in two-dimensional problems of geophysical fluid dynamics in the rotating fluid with random bottom topography (see, for example, papers [4–7]) and in the problem of anomalous structures on the sea surface (see Sect. 10.2).

Chapter 1
Two-Dimensional Geophysical Fluid Dynamics

1.1 Equilibrium Distributions for Hydrodynamic Flows

In the simplest case, the incompressible fluid flow in the two-dimensional plane $\mathbf{R} = (x, y)$ is described by the stream function $\psi(\mathbf{R}, t)$ satisfying the equation:

$$\frac{\partial}{\partial t} \Delta \psi(\mathbf{R}, t) = J \{\Delta \psi(\mathbf{R}, t); \psi(\mathbf{R}, t)\}, \quad \psi(\mathbf{R}, 0) = \psi_0(\mathbf{R}), \qquad (1.1)$$

where

$$J \{\psi(\mathbf{R}, t); \varphi(\mathbf{R}, t)\} = \frac{\partial \psi(\mathbf{R}, t)}{\partial x} \frac{\partial \varphi(\mathbf{R}, t)}{\partial y} - \frac{\partial \varphi(\mathbf{R}, t)}{\partial x} \frac{\partial \psi(\mathbf{R}, t)}{\partial y}$$

is the Jacobian of two functions [8].

Nonlinear interactions must bring the hydrodynamic system (1.1) to statistical equilibrium. In view of the fact that establishing this equilibrium requires a great number of interactions between the disturbances of different scales, we can suppose that, in the simplest case of statistically homogeneous and isotropic initial random field $\psi_0(\mathbf{R})$, this distribution will be the Gaussian distribution, so that our task consists in the determination of this distribution parameters. During the evolution, random stream function $\psi(\mathbf{R}, t)$ remains a homogeneous and isotropic function. Because the stream function is defined to an additive constant, we can describe its statistical characteristics by the one-time structure function

$$D_\psi(\mathbf{R} - \mathbf{R}', t) = \left\langle \left[\psi(\mathbf{R}, t) - \psi(\mathbf{R}', t)\right]^2 \right\rangle = 2 \left[B_\psi(0, t) - B_\psi(\mathbf{R} - \mathbf{R}', t) \right],$$

where

$$B_\psi(\mathbf{R} - \mathbf{R}', t) = \left\langle \psi(\mathbf{R}, t) \psi(\mathbf{R}', t) \right\rangle$$

is the spatial correlation function of field $\psi(\mathbf{R}, t)$.

© Springer International Publishing AG 2017
V.I. Klyatskin, *Fundamentals of Stochastic Nature Sciences*,
Understanding Complex Systems, DOI 10.1007/978-3-319-56922-2_1

We will seek the steady-state (equilibrium) distribution on the class of the Gaussian distributions of statistically homogeneous and isotropic field $\psi(\mathbf{R}, t)$ described by the structure function $D_\psi(R) = \lim_{t \to \infty} D_\psi(\mathbf{R}, t)$. With this goal in view, we consider the three-point equality

$$\frac{\partial}{\partial t} \langle \Delta\psi(\mathbf{R}_1, t) \Delta\psi(\mathbf{R}_2, t) \Delta\psi(\mathbf{R}_3, t) \rangle = 0$$

for $t \to \infty$ from which follows that

$$\frac{\partial}{\partial t} \langle \Delta\psi(\mathbf{R}_1, t) \Delta\psi(\mathbf{R}_2, t) \Delta\psi(\mathbf{R}_3, t) \rangle = \{1\} + \{2\} + \{3\} = 0, \qquad (1.2)$$

where by $\{1\}$ we designate the variable

$$\{1\} = \langle J\{\Delta\psi(\mathbf{R}_1, t); \psi(\mathbf{R}_1, t)\Delta\psi(\mathbf{R}_2, t)\Delta\psi(\mathbf{R}_3, t)\} \rangle, \qquad (1.3)$$

while the variables $\{2\}$ and $\{3\}$ correspond to cyclic permutation on the vectors $\{\mathbf{R}_1, \mathbf{R}_2, \text{ and } \mathbf{R}_3\}$.

Expression (1.3) can be rewritten in the form

$$\{1\} = \left\langle \frac{\partial \Delta\psi(\mathbf{R}_1, t)}{\partial x_1} \frac{\partial \psi(\mathbf{R}_1, t)}{\partial y_1} \Delta\psi(\mathbf{R}_2, t)\Delta\psi(\mathbf{R}_3, t) \right\rangle$$
$$- \left\langle \frac{\partial \Delta\psi(\mathbf{R}_1, t)}{\partial y_1} \frac{\partial \psi(\mathbf{R}_1, t)}{\partial x_1} \Delta\psi(\mathbf{R}_2, t)\Delta\psi(\mathbf{R}_3, t) \right\rangle.$$

Next we split the quantic correlation in Eq. (1.3) in the product of pair correlations using the Gaussian property of the field $\psi(\mathbf{R}, t)$

$$\{1\} = \left\langle \frac{\partial \Delta\psi(\mathbf{R}_1, t)}{\partial x_1} \Delta\psi(\mathbf{R}_2, t) \right\rangle \left\langle \frac{\partial \psi(\mathbf{R}_1, t)}{\partial y_1} \Delta\psi(\mathbf{R}_3, t) \right\rangle$$
$$+ \left\langle \frac{\partial \Delta\psi(\mathbf{R}_1, t)}{\partial x_1} \Delta\psi(\mathbf{R}_3, t) \right\rangle \left\langle \frac{\partial \psi(\mathbf{R}_1, t)}{\partial y_1} \Delta\psi(\mathbf{R}_2, t) \right\rangle$$
$$- \left\langle \frac{\partial \Delta\psi(\mathbf{R}_1, t)}{\partial y_1} \Delta\psi(\mathbf{R}_2, t) \right\rangle \left\langle \frac{\partial \psi(\mathbf{R}_1, t)}{\partial x_1} \Delta\psi(\mathbf{R}_3, t) \right\rangle$$
$$- \left\langle \frac{\partial \Delta\psi(\mathbf{R}_1, t)}{\partial y_1} \Delta\psi(\mathbf{R}_3, t) \right\rangle \left\langle \frac{\partial \psi(\mathbf{R}_1, t)}{\partial x_1} \Delta\psi(\mathbf{R}_2, t) \right\rangle. \qquad (1.4)$$

Here we dropped the product terms containing $\langle J\{\Delta\psi(\mathbf{R}, t); \psi(\mathbf{R}, t)\} \rangle = 0$ as it vanishes after the ensemble averaging. In addition, we can replace derivatives $\frac{\partial}{\partial \mathbf{R}_1}$ with $\left(\frac{\partial}{\partial \mathbf{R}_2}, \frac{\partial}{\partial \mathbf{R}_3} \right)$ and, consequently, express quantity $\{1\}$ in terms of the correlation function of current

$$\{1\} = \left(\frac{\partial^2}{\partial x_2 \partial y_3} - \frac{\partial^2}{\partial x_3 \partial y_2}\right) \Delta_{\mathbf{R}_2} \Delta_{\mathbf{R}_3} \left(\Delta_{\mathbf{R}_2} - \Delta_{\mathbf{R}_3}\right) B_\psi(\mathbf{R}_1 - \mathbf{R}_2) B_\psi(\mathbf{R}_1 - \mathbf{R}_3).$$

Introducing vectors

$$\mathbf{q}_1 = \mathbf{R}_1 - \mathbf{R}_2, \ \mathbf{q}_2 = \mathbf{R}_2 - \mathbf{R}_3, \ \mathbf{q}_3 = \mathbf{R}_3 - \mathbf{R}_1 \quad (\mathbf{q}_1 + \mathbf{q}_2 + \mathbf{q}_1 = 0)$$

and scalars $q_i = |\mathbf{q}_i|$ we replace all $\mathbf{R}$-partial derivatives through the $\mathbf{q}$-partial derivatives

$$\frac{\partial}{\partial x_2} = -\frac{x_1 - x_2}{q_1} \frac{\partial}{\partial q_1}, \quad \frac{\partial}{\partial y_2} = -\frac{y_1 - y_2}{q_1} \frac{\partial}{\partial q_1},$$

$$\frac{\partial}{\partial x_3} = -\frac{x_3 - x_1}{q_1} \frac{\partial}{\partial q_1}, \quad \frac{\partial}{\partial y_3} = -\frac{y_3 - y_1}{q_1} \frac{\partial}{\partial q_1}.$$

The off-shot is the following equation

$$\{1\} = -[\mathbf{q}_3 \times \mathbf{q}_1] X(q_3; q_1),$$

where we denote by $[\mathbf{q}_3 \times \mathbf{q}_1]$ the wedge-product of two vectors, and $X(q_3; q_1)$ is given by the expression

$$X(q_3; q_1) = \frac{1}{q_3 q_1} \frac{\partial^2}{\partial q_3 \partial q_1} \left\{\Delta_{q_3}^2 D_\psi(q_3) \Delta_{q_1} D_\psi(q_1) - \Delta_{q_3} D_\psi(q_3) \Delta_{q_1}^2 D_\psi(q_1)\right\},$$

$$(1.5)$$

where $D_\psi(q)$ is the structure function of current and operator $\Delta_q = \frac{\partial^2}{\partial q^2} + \frac{1}{q} \frac{\partial}{\partial q}$ is the radial part of the Laplace operator.

As a result, the fundamental equality (1.2) reduces to the expression

$$[\mathbf{q}_3 \times \mathbf{q}_1] X(q_3; q_1) + [\mathbf{q}_2 \times \mathbf{q}_3] X(q_2; q_3) + [\mathbf{q}_1 \times \mathbf{q}_2] X(q_1; q_2) = 0.$$

In view of equalities $[\mathbf{q}_3 \times \mathbf{q}_1] = [\mathbf{q}_2 \times \mathbf{q}_3] = [\mathbf{q}_1 \times \mathbf{q}_2]$, it grades into the final functional equation

$$X(q_1; q_2) + X(q_2; q_3) + X(q_3; q_1) = 0 \qquad (1.6)$$

in arbitrary scalar coordinates $q_1 = |\mathbf{R}_1 - \mathbf{R}_2|$, $q_2 = |\mathbf{R}_2 - \mathbf{R}_3|$, $q_3 = |\mathbf{R}_3 - \mathbf{R}_1|$.

Then, multiplying Eq. (1.6) by $q_1 q_2$ and applying the differential operator $\partial^4/\partial q_1^2 \partial q_2^2$ to the product, we can eliminate variable q_3 and convert it to the equation for function D_ψ,

$$\frac{\partial^6}{\partial q_1^3 \partial q_2^3} \left\{\Delta_{q_1}^2 D_\psi(q_1) \Delta_{q_2} D_\psi(q_2) - \Delta_{q_1} D_\psi(q_1) \Delta_{q_2}^2 D_\psi(q_2)\right\} = 0.$$

Assuming now that function $\Delta_q D_\psi(q) \to 0$ for $q \to \infty$, we obtain the equation

$$\Delta_{q_1}^2 D_\psi(q_1) \Delta_{q_2} D_\psi(q_2) - \Delta_{q_1} D_\psi(q_1) \Delta_{q_2}^2 D_\psi(q_2) = 0, \tag{1.7}$$

which can be solved by the method of separation of variables. As a result, we arrive at the equation of the form [4]

$$\left(\Delta_q + \lambda\right) \Delta_q D_\psi(q) = 0, \tag{1.8}$$

where λ is the separation constant with the dimension of the inverse square of length and Δ_q is the radial part of the Laplace operator.

There are two possible solutions to Eq. (1.8), depending on whether constant λ is positive ($\lambda = k_0^2 > 0$) or negative ($\lambda = -k_0^2 < 0$).

If $\lambda = k_0^2 > 0$, Eq. (1.8) can be reduced to the equation

$$\Delta_q D_\psi(q) = C J_0(k_0 q),$$

where $J_0(z)$ is the Bessel function of the first kind. In this case, structure function $D_\psi(q)$ is determined as the solution to the Laplace equation, and we obtain the spectral density of current in the form

$$E(k) = E\delta(k - k_0).$$

The delta-like behavior of spectral density is evidence of the fact that fields $\psi(\mathbf{R}, t)$ are highly correlated, which suggests that coherent structures can exist in the developed turbulent flow of the two-dimensional fluid (in the sense of the existence of the corresponding eigenfunctions slowly decaying with distance). This pattern corresponds to random structures characterized by certain **fixed spatial scale**. In the problem under consideration, such structures are vortices, which means that structure formation is realized here in the form of *vortex genesis*.

In the case $\lambda = -k_0^2 < 0$, Eq. (1.8) can be reduced to the similar equation

$$\Delta_q D_\psi(q) = C K_0(k_0 q).$$

However, the right-hand side of this equation is proportional to the McDonalds function $K_0(z)$ with the dimensional parameters k_0 and C. The corresponding spectral density of current has the form

$$E(k) = \frac{E_0}{k^2 + k_0^2},$$

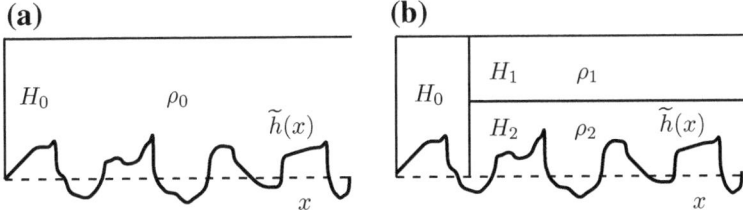

Fig. 1.1 Diagrammatic views of **a** one-layer and **b** two-layer models of hydrodynamic flows

which corresponds to the Gibbs distribution with two integrals of motion - those of energy and vorticity squared (enstrophy) of a velocity field (see, for example, Refs. [9, 10]).

Consider now the description of hydrodynamic flows on the rotating Earth in the so-called *quasi-geostrophic approximation*. In the simplest case of the one-layer model, the incompressible fluid flow in the two-dimensional plane $\mathbf{R} = (x, y)$ is described by the *stream function* that satisfies the equation

$$\frac{\partial}{\partial t} \Delta \psi(\mathbf{R}, t) + \beta_0 \frac{\partial}{\partial x} \psi(\mathbf{R}, t) = J \{\Delta \psi(\mathbf{R}, t) + h(\mathbf{R}); \psi(\mathbf{R}, t)\},$$

where parameter β_0 is the derivative of the local *Coriolis parameter* f_0 with respect to latitude and function $h(\mathbf{R}) = f_0 \widetilde{h}(\mathbf{R})/H_0$ is the deviation of bottom topography $\widetilde{h}(\mathbf{R})$ relative to its average thickness H_0 (Fig. 1.1a).

This equation describes the *barotropic* motion of a fluid. In the more general case of *baroclinic* motions, investigation is usually carried out within the framework of the two-layer model of hydrodynamic flows described by the system of equations

$$\frac{\partial}{\partial t} [\Delta \psi_1 - \alpha_1 F (\psi_1 - \psi_2)] + \beta_0 \frac{\partial}{\partial x} \psi_1 = J \{\Delta \psi_1 - \alpha_1 F (\psi_1 - \psi_2); \psi_1\},$$
$$\frac{\partial}{\partial t} [\Delta \psi_2 - \alpha_2 F (\psi_2 - \psi_1)] + \beta_0 \frac{\partial}{\partial x} \psi_2 = J \{\Delta \psi_2 - \alpha_2 F (\psi_2 - \psi_1) + f_0 \alpha_2 h; \psi_2\},$$

where additional parameters

$$F = f_0^2 \rho/g(\Delta \rho), \quad \Delta \rho/\rho = (\rho_2 - \rho_1)/\rho_0 > 0$$

are introduced and $\alpha_1 = 1/H_1$ and $\alpha_2 = 1/H_2$ are the inverse thicknesses of layers (Fig. 1.1b).

Thus, there are already *two fixed scales* in a two-layer fluid.

(a)   **(b)**

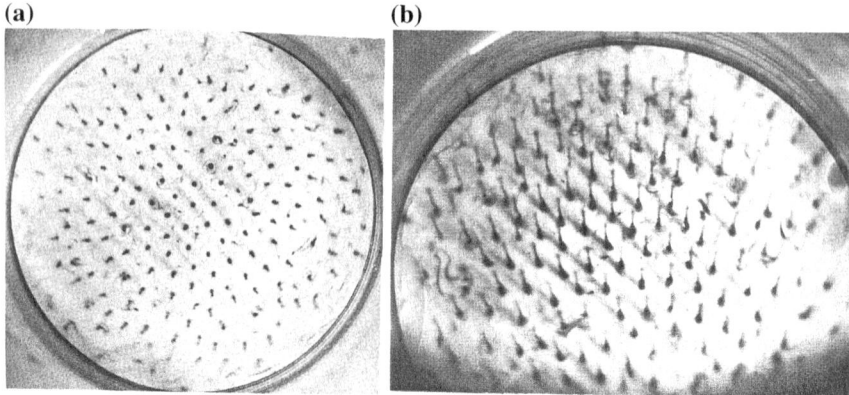

Fig. 1.2 System of regular converting vortices on a rotating platform: **a** *top view*, and **b** *side view* (Taken from monograph [15])

Fig. 1.3 System of irregular convective vortices on a rotating platform for a rotational velocity larger than in Fig. 1.2 (Taken from monograph [15])

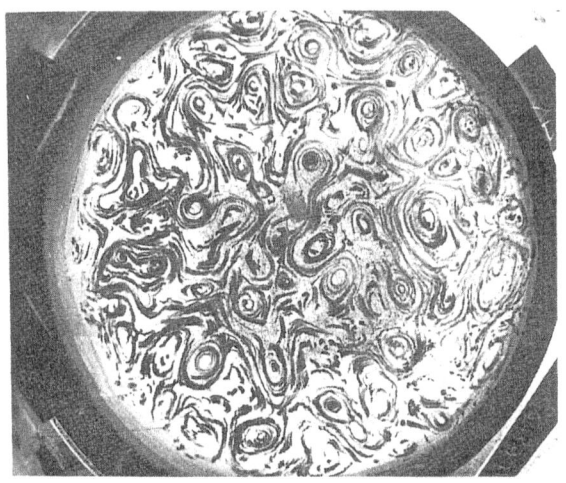

It seems plausible to assume that such structures have been observed in experimental studies in rotating fluids (see, for example, Refs. [11–13], review [14], and monographs [15, 16]), and also in numerical simulations (see, for example, Ref. [17]). As illustrations, we present Figs. 1.2, 1.3, 1.4, 1.5 which, in our opinion, correspond to the situation described. Figure 1.6 shows an example of structure formation in the field of surface flows in the Baltic Sea [18, 19].

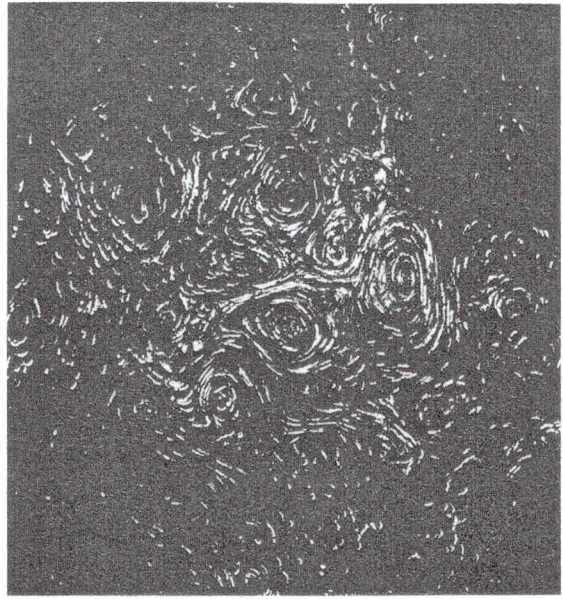

Fig. 1.4 Streak pattern of velocities in a baroclinically unstable two-layer stratified fluid on a rotating platform (Taken from Ref. [14])

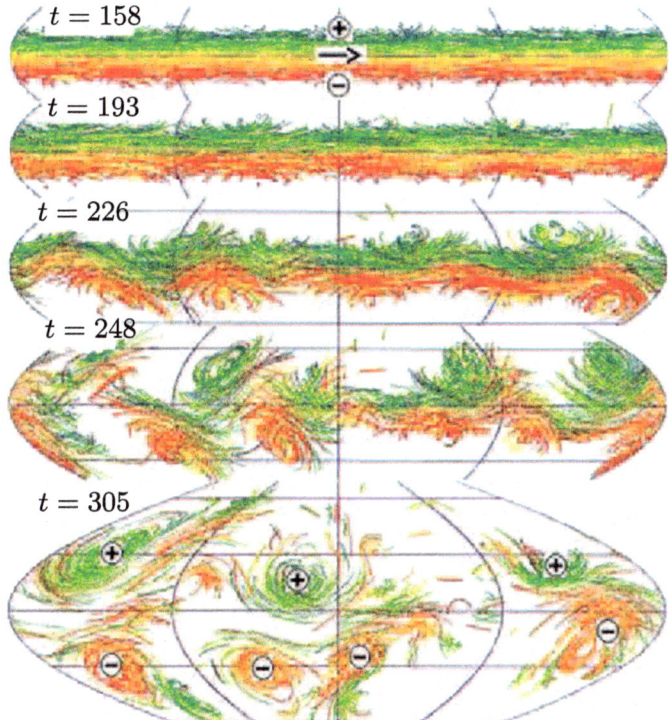

Fig. 1.5 Formation of eddies on a sphere driven by an unstable shear flow (Taken from Ref. [17])

Fig. 1.6 Domain of the
submesoscale surface
vorticity field in the Baltic
Sea

1.2 Plane Motion Under the Action of a Periodic Force

Consider now the two-dimensional motion of an incompressible viscous fluid
$\mathbf{U}(\mathbf{r}, t) = \{u(\mathbf{r}, t),\ v(\mathbf{r}, t)\}$ in plane $\mathbf{r} = \{x,\ y\}$ under the action of a spatially periodic
force directed along the x-axis,

$$f_x(\mathbf{r}, t) = \gamma \sin py \ (\gamma > 0).$$

Such a flow is usually called the *Kolmogorov flow* (*stream*). The corresponding
motion is described by the system of equations

$$\frac{\partial u}{\partial t} + \frac{\partial u^2}{\partial x} + \frac{\partial uv}{\partial y} = -\frac{1}{\rho}\frac{\partial P}{\partial x} + \nu \Delta u + \gamma \sin py,$$

$$\frac{\partial v}{\partial t} + \frac{\partial uv}{\partial x} + \frac{\partial v^2}{\partial y} = -\frac{1}{\rho}\frac{\partial P}{\partial y} + \nu \Delta v, \qquad (1.9)$$

$$\frac{\partial u}{\partial x} + \frac{\partial v}{\partial y} = 0,$$

where $P(\mathbf{r}, t)$ is the pressure, ρ is the density, and ν is the kinematic viscosity. See
papers [20, 21] for the laboratory modeling of the Kolmogorov flow.

The system of the Navier–Stokes and continuity equations (1.9) has the steady-
state solution that corresponds to the laminar flow at constant pressure along the
x-axis and has the following form

$$u_{\text{s-s}}(\mathbf{r}) = \frac{\gamma}{\nu p^2} \sin py, \quad v_{\text{s-s}}(\mathbf{r}) = 0, \quad P_{\text{s-s}}(\mathbf{r}) = \text{const}. \qquad (1.10)$$

Introducing scales of length p^{-1}, velocity $p^{-2}\nu^{-1}\gamma$, and time $p\nu\gamma^{-1}$ and using dimensionless variables, we reduce system (1.9) to the form

$$\frac{\partial u}{\partial t} + \frac{\partial u^2}{\partial x} + \frac{\partial uv}{\partial y} = -\frac{\partial P}{\partial x} + \frac{1}{R}\Delta u + \frac{1}{R}\sin y,$$

$$\frac{\partial v}{\partial t} + \frac{\partial uv}{\partial x} + \frac{\partial v^2}{\partial y} = -\frac{\partial P}{\partial y} + \frac{1}{R}\Delta v, \qquad (1.11)$$

$$\frac{\partial u}{\partial x} + \frac{\partial v}{\partial y} = 0,$$

where $R = \dfrac{\gamma}{\nu^2 p^3}$ is the Reynolds number. In these variables, the steady-state solution has the form

$$u_{\text{s-s}}(\mathbf{r}) = \sin y, \quad v_{\text{s-s}}(\mathbf{r}) = 0, \quad P_{\text{s-s}}(\mathbf{r}) = \text{const.}$$

Introducing flow function $\psi(\mathbf{r}, t)$ by the relationship

$$u(\mathbf{r}, t) = \frac{\partial}{\partial y}\psi(\mathbf{r}, t), \quad v(\mathbf{r}, t) = -\frac{\partial}{\partial x}\psi(\mathbf{r}, t),$$

we obtain that it satisfies the equation

$$\left(\frac{\partial}{\partial t} - \frac{\Delta}{R}\right)\Delta\psi - \frac{\partial\psi}{\partial x}\frac{\partial\Delta\psi}{\partial y} + \frac{\partial\psi}{\partial y}\frac{\partial\Delta\psi}{\partial x} = \frac{1}{R}\cos y \qquad (1.12)$$

and

$$\psi_{\text{s-s}}(\mathbf{r}) = -\cos y.$$

It was shown [22, 23] that, in the linear problem formulation, the steady-state solution (1.10) corresponding to the laminar flow is unstable with respect to small disturbances for certain values of parameter R. These disturbances rapidly increase in time getting the energy from the flow (1.10); this causes the Reynolds stresses described by the nonlinear terms in Eq. (1.12) to increase, which results in decreasing the amplitude of laminar flow until certain new steady-state flow (called usually the *secondary flow*) is formed.

Represent the hydrodynamic fields in the form

$$u(\mathbf{r}, t) = U(y, t) + \tilde{u}(\mathbf{r}, t), \quad v(\mathbf{r}, t) = \tilde{v}(\mathbf{r}, t),$$
$$P(\mathbf{r}, t) = P_0 + \tilde{P}(\mathbf{r}, t), \quad \psi(\mathbf{r}, t) = \Psi(y, t) + \tilde{\psi}(\mathbf{r}, t).$$

Here, $U(y, t)$ is the new profile of the steady-state flow to be determined together with the Reynolds stresses and the tilde denotes the corresponding ultimate disturbances.

Abiding by the cited works, we will consider disturbances harmonic in variable x with wavelength $2\pi/\alpha$ ($\alpha > 0$). The new flow profile $U(y, t)$ is the result of averaging with respect to x over distances of about a wavelength.

One can easily see that, for $\alpha \geq 1$, the laminar flow (1.10) is unique and stable for all R [23] and instability can appear only for disturbances with $\alpha < 1$.

According to the linear theory of stability, we will consider first the nonlinear interaction only between the first harmonic of disturbances and the mean flow and neglect the generation of higher harmonics and their mutual interactions and interaction with the mean flow.

We represent all disturbances in the form

$$
\begin{aligned}
\widetilde{\varphi}(\mathbf{r}, t) &= \varphi^{(1)}(y, t)e^{i\alpha x} + \varphi^{(-1)}(y, t)e^{-i\alpha x}, \\
\left(\widetilde{\varphi}(\mathbf{r}, t) \right. &= \widetilde{u}(\mathbf{r}, t), \ \widetilde{v}(\mathbf{r}, t), \ \widetilde{P}(\mathbf{r}, t), \ \widetilde{\psi}(\mathbf{r}, t)\big),
\end{aligned}
$$

where quantity $\varphi^{(-1)}(y, t)$ is the complex conjugated of $\varphi^{(1)}(y, t)$. Then, using this representation in system (1.11) and eliminating quantities $\widetilde{P}(\mathbf{r}, t)$ and $\widetilde{u}(\mathbf{r}, t)$, we obtain the system of equations in mean flow $U(y, t)$ and disturbances $v^{(1)}(y, t)$ [24, 26]:

$$
\frac{\partial}{\partial t} U + \frac{i}{\alpha} \left(v^{(-1)} \frac{\partial^2 v^{(1)}}{\partial y^2} - v^{(1)} \frac{\partial^2 v^{(-1)}}{\partial y^2} \right) = \frac{1}{R} \frac{\partial^2 U}{\partial y^2} + \frac{1}{R} \sin y,
$$

$$
(1.13)
$$

$$
\left(\frac{\partial}{\partial t} - \frac{\Delta}{R} \right) \Delta v^{(1)} + i\alpha \left[U \Delta v^{(1)} - v^{(1)} \frac{\partial^2 U}{\partial y^2} \right] = 0.
$$

The second equation in system (1.13) is the known *Orr–Sommerfeld equation*. A similar system can be derived for the flow function.

To examine the stability of the laminar regime (1.10), we set

$$
U(y) = \sin y
$$

in the second equation of system (1.13) to obtain

$$
\left(\frac{\partial}{\partial t} - \frac{\Delta}{R} \right) \Delta v^{(1)}(y, t) + i\alpha \sin y [1 + \Delta] v^{(1)}(y, t) = 0. \qquad (1.14)
$$

Representing disturbances $v^{(1)}(y, t)$ in the form

$$
v^{(1)}(y, t) = \sum_{n=-\infty}^{\infty} v_n^{(1)} e^{\sigma t + iny} \qquad (1.15)
$$

and substituting this representation in Eq. (1.14), we arrive at the recurrent system in quantities $v_n^{(1)}$

$$\frac{2}{\alpha} \left(\alpha^2 + n^2\right) \left[\sigma + \frac{\alpha^2 + n^2}{R}\right] v_n^{(1)} + v_{n-1}^{(1)} \left[\alpha^2 - 1 + (n - 1)^2\right]$$
$$- v_{n+1}^{(1)} \left[\alpha^2 - 1 + (n + 1)^2\right] = 0, \qquad n = -\infty, \cdots, +\infty. \quad (1.16)$$

The analysis of system (1.16) showed [22, 23] that, under certain restrictions on wave number α and Reynolds number R, positive values of parameter σ can exist, i.e., the solutions are unstable. The corresponding dispersion equation in σ has the form of an infinite continued fraction, and the critical Reynolds number is $R_{\mathrm{cr}} = \sqrt{2}$ for $\alpha \to 0$. In other words, long-wave disturbances along the applied force appear to be most unstable. For this reason, we can consider parameter α as a small parameter of the problem at hand and integrate the Orr–Sommerfeld equation asymptotically. We will not dwell on details of this solution. Note only that the components of eigenvector $\{v_n^{(1)}\}$ of problem (1.16) have different orders of magnitude in parameter α. For example, all components of vector $\{v_n^{(1)}\}$ with $n = \pm 2, \pm 3, \cdots$ will have an order of α^4 at least. As a result, we can confine ourselves to the most significant harmonics with $n = 0, \pm 1$, which is, in essence, equivalent to the *Galerkin method* with the trigonometric coordinate functions. In this case,

$$U(y, t) = U(t) \sin y,$$

and the equation in $v^{(1)}$ assumes the form

$$\left(\frac{\partial}{\partial t} - \frac{\Delta}{R}\right) \Delta v^{(1)}(y, t) + i\alpha U(t) \sin y [1 + \Delta] v^{(1)}(y, t) = 0.$$

Substituting the expansion

$$v^{(1)}(y, t) = \sum_{n=-1}^{1} v_n^{(1)}(t) e^{iny}$$

in Eq. (1.13), we obtain that functions

$$U(t), \quad z_0(t) = v_0^{(1)}(t), \quad z_+(t) = v_1^{(1)}(t) + v_{-1}^{(1)}(t),$$

$$z_-(t) = \frac{v_1^{(1)}(t) - v_{-1}^{(1)}(t)}{2}$$

satisfy the system of equations [24, 26]

$$\left(\frac{d}{dt} + \frac{1}{R}\right) U(t) = \frac{1}{R} - \frac{4}{\alpha} z_0(t) z_-(t),$$

$$\left(\frac{d}{dt} + \frac{\alpha^2}{R}\right) z_0(t) = \alpha U(t) z_-(t), \tag{1.17}$$

$$\left(\frac{d}{dt} + \frac{1}{R}\right) z_-(t) = \frac{\alpha}{2} U(t) z_0(t), \quad \left(\frac{d}{dt} + \frac{1}{R}\right) z_+(t) = 0.$$

Equation in quantity $z_+(t)$ is independent of other equations; as a consequence, the corresponding disturbances can only decay with time. Three remaining equations form the simplest three-component hydrodynamic-type system (see monograph [24]). As was mentioned earlier, this system is equivalent to the dynamic system describing the motion of a gyroscope with anisotropic friction under the action of an external moment of force relative to the unstable axis. An analysis of system (1.17) shows that, for $R < R_{cr} = \sqrt{2}$, it yields the laminar regime with $U = 1, z_i = 0$. For $R > \sqrt{2}$, this regime becomes unstable, and new regime (the secondary flow) is formed that corresponds to the mean flow profile and steady-state Reynolds stresses

$$U = \frac{\sqrt{2}}{R}, \quad \frac{4}{\alpha} z_0 z_- = \frac{R - \sqrt{2}}{R^2}, \quad z_+ = 0,$$

$$\left[v_0^{(1)}\right]^2 = \frac{R - \sqrt{2}}{2\sqrt{2} R^2}, \quad v_1^{(1)} = \frac{\alpha}{\sqrt{2}} v_0^{(1)}, \quad \alpha \ll 1, \quad R \geq \sqrt{2}.$$

Turning back to the dimensional quantities, we obtain

$$U(y) = \sqrt{2} v p \sin py, \quad \langle \widetilde{u}\widetilde{v} \rangle = -\frac{\gamma}{p} \frac{R - \sqrt{2}}{R} \cos py. \tag{1.18}$$

Note that the amplitude of the steady-state mean flow is independent of the amplitude of exciting force. Moreover, quantity $v_0^{(1)}$ can be both positive and negative, depending on the signs of the amplitudes of small initial disturbances.

Flow function of the steady-state flow has the form

$$\psi_1(x, y) = -\frac{\sqrt{2}}{R} \cos y - \frac{2}{\alpha} v_0^{(1)} \left[\sqrt{2}\alpha \sin y \cos \alpha x + \sin \alpha x\right].$$

Figure 1.7 shows the current lines

$$\alpha \cos y + \sqrt{2}\alpha \sin y \cos \alpha x + \sin \alpha x = C$$

of flow (1.18) at $R = 2R_{cr} = 2\sqrt{2}$ ($v_0^{(1)} > 0$). In addition, Fig. 1.7 shows schematically the profile of the mean flow. As distinct from the laminar solution, systems of

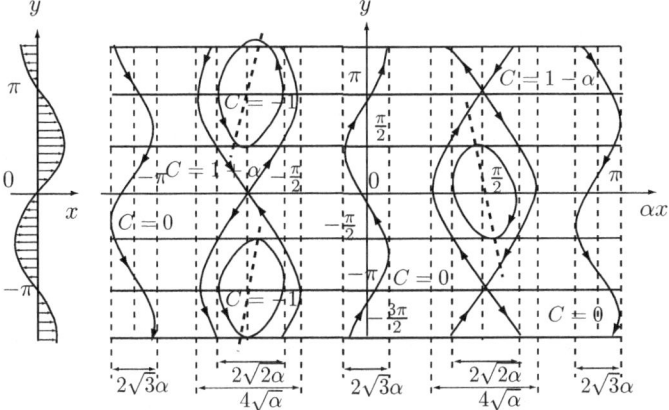

Fig. 1.7 Mean velocity and current lines of the secondary flow at $R = 2R_{cr} = 2\sqrt{2}$ ($\tilde{v}_0 > 0$)

spatially periodic vortices appear here, and the tilt of longer axes of these vortices is determined by the sign of the derivative of the mean flow profile with respect to y.

Flow (1.18) was derived under the assumption that the nonlinear interactions between different harmonics of the disturbance are insignificant in comparison with their interactions with the mean flow. This assumption will hold if flow (1.18) is, in turn, stable with respect to small disturbances. The corresponding stability analysis can be carried out by the standard procedure, i.e., by linearizing the equation for flow function (1.12) relative to flow (1.18) [26]. The analysis shows that flow (1.18) is stable if we restrict ourselves to the harmonics of the types same as in solution (1.18).

However, the solution appears to be unstable with respect to small-scale disturbances. In this case, the nonlinear interaction of infinitely small disturbances governs the motion along with the interaction of the disturbances with the mean flow. Moreover, we cannot here content ourselves with a finite number of harmonics in x-coordinate and need to consider the infinite series. As regards the harmonics in y-coordinate, we, as earlier, can limit the consideration to the harmonics with $n = 0, \pm 1$.

Represent the flow function in the form

$$\psi(x, y, t) = \psi_{-1}(x, t)e^{-iy} + \psi_0(x, t) + \psi_1(x, t)e^{iy}$$

(1.19)

$$(\psi_1^*(x, t) = \psi_{-1}(x, t)),$$

where $\psi_i(x, t)$ are the periodic functions in x-coordinate with a period $2\pi/\alpha$. Substituting Eq. (1.19) in Eq. (1.12), neglecting the terms of order α^3 in interactions of harmonics and the terms of order α^2 in the dissipative terms of harmonics $\psi_{\pm 1}$, and introducing new functions

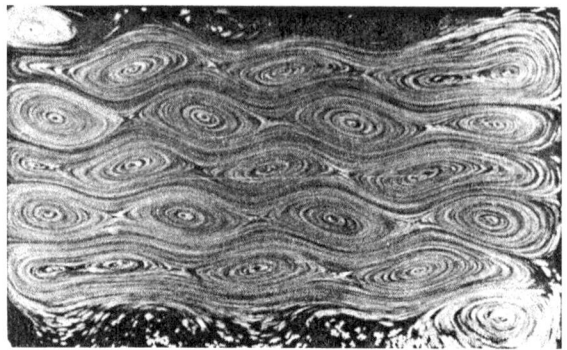

$$\psi_+(x,t) = \frac{\psi_{-1} + \psi_1}{2}, \quad \psi_-(x,t) = \frac{\psi_{-1} - \psi_1}{2i},$$

we obtain the system of equations

$$\left(\frac{\partial}{\partial t} + \frac{1}{R}\right)\psi_+ - \psi_-\frac{\partial\psi_0}{\partial x} = -\frac{1}{2R},$$

$$\left(\frac{\partial}{\partial t} + \frac{1}{R}\right)\psi_- + \psi_+\frac{\partial\psi_0}{\partial x} = 0,$$

$$\left(\frac{\partial}{\partial t} - \frac{1}{R}\frac{\partial^2}{\partial x^2}\right)\psi_0 + 2\left(\psi_-\frac{\partial\psi_+}{\partial x} - \psi_+\frac{\partial\psi_-}{\partial x}\right) = 0.$$

This system of equations takes into account the whole infinite series of harmonics in x-coordinate and extends the system of gyroscopic-type equations (1.17) to the infinite-dimensional case. Its characteristic feature consists in the absence of steady-state solutions periodic in x-coordinate (except the solution corresponding to the laminar flow).

Figure 1.8 shows an example of the Kolmogorov flow obtained experimentally (see, e.g., [27]).

Chapter 2
Parametrically Excited Dynamic Systems

We turn now to the statistical analysis of stochastic dynamic systems related to random parametric excitation in space and time. Such systems, appearing in many branches of physics, can be described by ordinary differential equations, as well as by partial differential equations. Stochastic structure formation for such systems in random media in the form of *clustering* is related to the parametric excitation of various physical fields in these media. *Clustering of a particular field implies the appearance of compact regions with large field values against the background of surrounding areas with relatively low field values.* Statistical averaging, expectedly, destroys all the information on clusters. Such challenges occur in fluid dynamics (*a passive scalar tracer in a turbulent flow*), in magnetohydrodynamics (*a passive vector tracer—magnetic field in a turbulent flow*), and in the propagation of waves of various origins (acoustic and radio waves, light and laser radiation) in random media. All these issues are commonly considered in the kinematic approximation and share the following two most important traits.

1. At fixed points in space, the field realizations in time are random processes which possess a specific character: they have the shape of peaks that appear at random instants of time. The intervals between them are characterized by low intensity and long duration. Such a realization of a random process in time for any location in space stems from the lognormal one-time distribution of probabilities, which has a slightly sloping 'tail'. The large but rare outliers (fluctuations) come from these tails. The main statistical characteristics of the processes being considered are the one-time probability density, one-time moment functions, typical realization curve characterizing the key features in the behavior of realizations of random processes, and Lyapunov exponent. In one-dimensional tasks described by ordinary differential equations with initial or boundary conditions, only such physical phenomena as *dynamic localization* can be observed in a number of cases (see Chap. 5).

2. The structure formation itself of a stochastic field takes place in physical space and is described through a related statistical analysis based on the ideas of statistical topography of a stochastic field. In the simplest problem formulation, under

© Springer International Publishing AG 2017

V.I. Klyatskin, *Fundamentals of Stochastic Nature Sciences*,
Understanding Complex Systems, DOI 10.1007/978-3-319-56922-2_2

statistical homogeneity in space, all one-point statistical characteristics of a random field are independent of spatial locations. Accordingly, the equation for the one-point probability density of a random field coincides in form with the equation for the probability density of a random process at each point in space, although the sense of these equations is substantially different. Accordingly, the statistical analysis of these equations should also be completely different.

A detailed discussion of these questions can be found in monographs [28, 29] and articles [30–33, 35].

First of all, a question arises as to whether or not such physical phenomena as localization and clustering appear in individual realizations of the processes and fields being considered, and if yes, then over which characteristic time (or on which spatial scales).

2.1 Lognormal Random Process

The phenomenon of structure formation in stochastic, parametrically excited dynamic systems on its own is well known in physics. For example, solutions of one-dimensional problems on parametric excitation, described by ordinary differential equations, are random processes.

The simplest dynamic system of that kind defines a lognormal random process $y(t; \alpha)$ described by a first-order ordinary stochastic differential equation:

$$\frac{d}{dt} y(t; \alpha) = \{-\alpha + z(t)\} y(t; \alpha), \quad y(0; \alpha) = 1, \tag{2.1}$$

where $z(t)$ is a Gaussian random process of *white noise* with the parameters

$$\langle z(t) \rangle = 0, \quad B_z(t - t') = \langle z(t) z(t') \rangle = 2D\delta(t - t').$$

The solution to Eq. (2.1) takes the form

$$y(t; \alpha) = \exp \left\{ -\alpha t + \int_0^t d\tau z(\tau) \right\}. \tag{2.2}$$

It should be noted that the change in the sign of parameter α in Eq. (2.2) is statistically equivalent to passing to the process $1/y(t)$ [36].

Figure 2.1 presents realizations of the lognormal random process $y(t; \alpha)$ given by formula (2.2) for positive and negative values of the parameter α and $|\alpha|/D = 1$ (the dashed lines correspond to the functions $\exp\{-Dt\}$ for $\alpha > 0$, and $\exp\{Dt\}$ for $\alpha < 0$). The presence of rare, but strong, spikes (fluctuations) with respect to the dashed curves in the directions of both large and small values can be seen in Fig. 2.1.

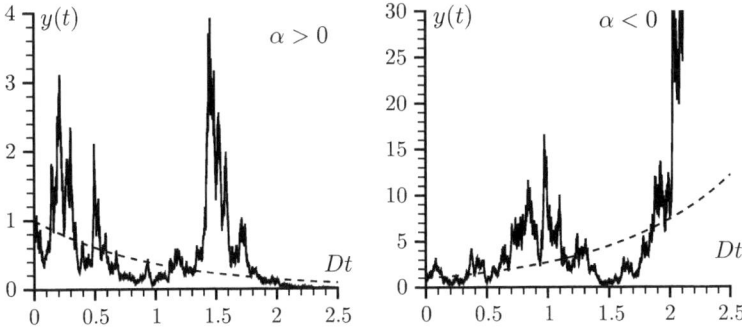

Fig. 2.1 Realizations of the lognormal process $y(t)$ for $\alpha > 0$, $\alpha < 0$ and $|\alpha|/D = 1$

This property of random processes is called *intermittency*; it was intensively studied in the 1980s (see, for example, Refs. [37, 38]). The curve with respect to which we identify the outliers (fluctuations) will be referred to as the *typical realization curve*.

2.2 Uncorrected Error of the Past Unavoidably Results in Errors of the Present and Future

> Once you have missed the first buttonhole,
> you will never manage to button up.
>
> Johann Wolfgang von Goete

The authors of practically every one of the numerous articles exploring the properties of intermittency cite paper [38] when turning to the notion of 'intermittency'. The term *intermittency* on its own emerged in studies of the velocity field and temperature spots in turbulent media [39, 40] (as, for instance, stated in Ref. [38]). However, even at that time it was already well known that one-point distributions of velocity fields and temperature fluctuations are close to the Gaussian ones (see, for example, Ref. [41]). The term *intermittency* is certainly a telling one, and it characterizes temporal variability of a random field at a fixed location in space, i.e., the variability of a random process with respect to its mean value.

At that time, certainly, it was also known that stochastic instability (parametric excitation) could occur in dynamic systems as a consequence of fluctuations in the internal parameters of the system. However, for a long time, up to the 1980s, nobody took interest in these questions. The merit of the authors of Ref. [38] is that they, in all probability, were the first to draw attention to the possibility of stochastic structure formation as a consequence of such parametric excitation, which had been known at that time from various kinds of observations.

The abstract to paper [38] states, "The processes of instability in random media are characterized by formation of specific structures in which a growing quantity reaches record-high values. Despite the rareness of such concentrations namely they confine the main part of integral characteristics of a growing quantity (the mean value, the mean squared value, etc.). The appearance of such structures is referred to as the phenomenon of *intermittency*."

Further, on page 2 of this paper, one can read, 'Structures appeared in a random medium show peculiar features; they have the form of spikes (outliers) appeared at random locations and at random times. The inter-spike intervals are characterized by low intensity and long extent. The general name of such a pattern is *intermittency*'.

Thus, in Ref. [38] strong rare outliers (fluctuations) are termed *specific structures*, while the process proper whereby these structures (outliers or fluctuations) are formed is called the phenomenon of *intermittency*.

In our understanding, *intermittency* is a general property of all random processes, independent of the amplitude of possible fluctuations, while *structure formation constitutes a certain type of evolution of stochastic dynamic systems in space and time*.

Reference [38] treated these notions as identical. At present, for example, some scientists call large rare outliers (fluctuations) characteristic of both the stochastic linear and nonlinear Leontovich equations (see below) rogue (freak) waves (see, for example, the lectures by Zakharov [42], paper [43] and monograph [44]). A rogue wave is undoubtedly a phenomenon of spatio-temporal watermass clustering, and it should be considered on the basis of an appropriate statistical analysis of the evolution of random fields.

At the same time, p. 5 of paper [38] has the following text: 'The main sign of intermittency is namely an anomalous relationship between successive statistical moments (in comparison with the Gaussian one)'. Thus, this paper opposes the Gaussian random processes and fields to the parametrically excited ones. However, as we have seen in Sect. 1.1, stochastic structure formation can occur even in the Gaussian fields.

It should be noted that the statistical theory of stationary extremal statistical processes is an independent branch of probability theory (see, for example, review [45]). However, in our opinion, this area has nothing to do with stochastic structure formation in space and time.

A fundamental feature of stochastic dynamic systems described by partial differential equations is that their solutions comprise random fields in space and time. The difficulty in explaining processes of structure formation in these systems is related to two factors. First, at any fixed point in space, the random field constitutes a random process in time. Second, for any fixed instant of time, the random field represents a random process over its spatial coordinates. Intermittency (i.e., variability) occurs namely for random processes (with respect to time or spatial coordinate); it is a general property of any random process irrespective of the *nature of its origin*.

In this study, the intermittency of a random process is understood as a more or less uninterrupted alteration of outliers (fluctuations) of this process toward larger as well as lower values with respect to the deterministic curve—*the curve of typical realization*, which is the median of the integral probability distribution function (see Sect. 5.1). In this case, a lognormal, parametrically excited random process can exponentially decay with time in individual realizations (certainly, with some fluctuations), which corresponds to the phenomenon of *dynamic localization*. An exponential growth of a random process with time is also possible, which corresponds to the absence of dynamic localization. A peculiarity of a lognormal random process is the presence of rare anomalously high spikes (fluctuations) on the curves of the process (see Fig. 2.1), related to the long sloped 'tail' of the probability density (see Chap. 5). All traditional statistical characteristics, such as moment and correlation functions of arbitrary order, result from these fluctuations.

By introducing the notion of the typical realization curve for a random process, we return to the historical sense of the concept of intermittency, which is general for all random processes and has a strict probabilistic definition and a transparent physical meaning.

2.3 Oscillator with Randomly Varying Frequency (Stochastic Parametric Resonance)

A more complicated problem that cannot be solved in analytical form is the problem on *stochastic parametric resonance*. Such a system is described by the second-order equation

$$\frac{d^2}{dt^2}x(t) + \omega_0^2[1 + z(t)]x(t) = 0,$$
$$x(0) = x_0, \quad \frac{d}{dt}x(0) = v_0, \tag{2.3}$$

where $z(t)$ is the random function of time. This equation is characteristic of almost all fields of physics. It is physically obvious that dynamic system (2.3) is capable of parametric excitation, because random process $z(t)$ has harmonic components of all frequencies, including frequencies $2\omega_0/n(n = 1, 2, \ldots)$ that exactly correspond to the frequencies of parametric resonance in the system with periodic function $z(t)$, as, for example, in the case of the *Mathieu equation*.

This problem will be statistically analyzed in Chap. 6.

Chapter 3
Examples of Stochastic Dynamic Systems

3.1 Particles Under the Random Velocity Field

One of the simplest physical problems related to parametric excitation is the problem on diffusion of a particle or an ensemble of particles in a random velocity field $\mathbf{u}(\mathbf{r}, t)$ with given statistical properties in the kinematic approximation (see, for example, monographs [28, 29], which provide an extensive bibliography of problems considered). The problem is described by the system of ordinary differential equations of the first order

$$\frac{d}{dt}\mathbf{r}(t) = \mathbf{u}(\mathbf{r}(t), t), \quad \mathbf{r}(0) = \mathbf{r}_0. \tag{3.1}$$

From Eq. (3.1) formally follows that every particle moves independently of other particles. However, if random field $\mathbf{u}(\mathbf{r}, t)$ has a finite spatial correlation radius l_{cor}, particles spaced by a distance shorter than l_{cor} appear in the common zone of infection of random field $\mathbf{u}(\mathbf{r}, t)$ and the behavior of such a system can show new collective features.

For steady velocity field $\mathbf{u}(\mathbf{r}, t) \equiv \mathbf{u}(\mathbf{r})$, Eq. (3.1) reduces to

$$\frac{d}{dt}\mathbf{r}(t) = \mathbf{u}(\mathbf{r}), \quad \mathbf{r}(0) = \mathbf{r}_0. \tag{3.2}$$

This equation clearly shows that steady points $\widetilde{\mathbf{r}}$ (at which $\mathbf{u}(\widetilde{\mathbf{r}}) = 0$) remain the fixed points. Depending on whether these points are stable or unstable, they will attract or repel nearby particles. In view of randomness of function $\mathbf{u}(\mathbf{r})$, points $\widetilde{\mathbf{r}}$ are random too.

It is expected that the similar behavior will be also characteristic of the general case of space-time random velocity field $\mathbf{u}(\mathbf{r}, t)$.

If some points $\widetilde{\mathbf{r}}$ remain stable during sufficiently long time, then clusters of particles (i.e., compact regions with elevated particle concentration, which occur merely in rarefied zones) must arise around these points in separate realizations of random field $\mathbf{u}(\mathbf{r}, t)$. On the contrary, if the stability of these points alternates with

© Springer International Publishing AG 2017
V.I. Klyatskin, *Fundamentals of Stochastic Nature Sciences*,
Understanding Complex Systems, DOI 10.1007/978-3-319-56922-2_3

instability sufficiently rapidly and particles have no time for significant rearrangement, no clusters of particles will occur.

Simulations show that the behavior of a system of particles essentially depends on whether the random field of velocities is nondivergent or divergent. By way of example, Fig. 3.1a shows a schematic of the evolution of the two-dimensional system of particles uniformly distributed within the circle for a particular realization of the nondivergent steady field $\mathbf{u}(\mathbf{r})$.

Here, we use the dimensionless time related to statistical parameters of field $\mathbf{u}(\mathbf{r})$. In this case, the area of surface patch within the contour remains intact and particles relatively uniformly fill the region within the deformed contour. The only feature consists in the fractal-type irregularity of the deformed contour. This phenomenon—called *chaotic advection*—is under active study now.

On the contrary, in the case of the potential velocity field $\mathbf{u}(\mathbf{r})$, particles uniformly distributed in the square at the initial instant will form clusters during the temporal evolution. Results simulated for this case are shown in Fig. 3.1b. We emphasize that the formation of clusters is purely a kinematic effect. This feature of particle dynamics disappears on averaging over an ensemble of realizations of random velocity field.

3.1.1   The Simplest Numerical Example of Particle Dynamics

To demonstrate the process of particle clustering we consider the simplest problem [71], in which random velocity field $\mathbf{u}(\mathbf{r}, t)$ has the form

$$\mathbf{u}(\mathbf{r}, t) = \mathbf{v}(t) f(\mathbf{kr}), \tag{3.3}$$

where $\mathbf{v}(t)$ is the random vector process and

$$f(\mathbf{kr}) = \sin(2\mathbf{kr}) \tag{3.4}$$

is the deterministic function of one variable. This form of function $f(\mathbf{kr})$ corresponds to the first term of the expansion in harmonic components and is commonly used in numerical simulations.

In this case, Eq. (3.1) can be written in the form

$$\frac{d}{dt}\mathbf{r}(t) = \mathbf{v}(t) \sin(2\mathbf{kr}), \quad \mathbf{r}(0) = \mathbf{r}_0.$$

In the context of this model, motions of a particle along vector $\mathbf{k}$ and in the plane perpendicular to vector $\mathbf{k}$ are independent and can be separated. If we direct the x-axis along vector $\mathbf{k}$, then the equations assume the form

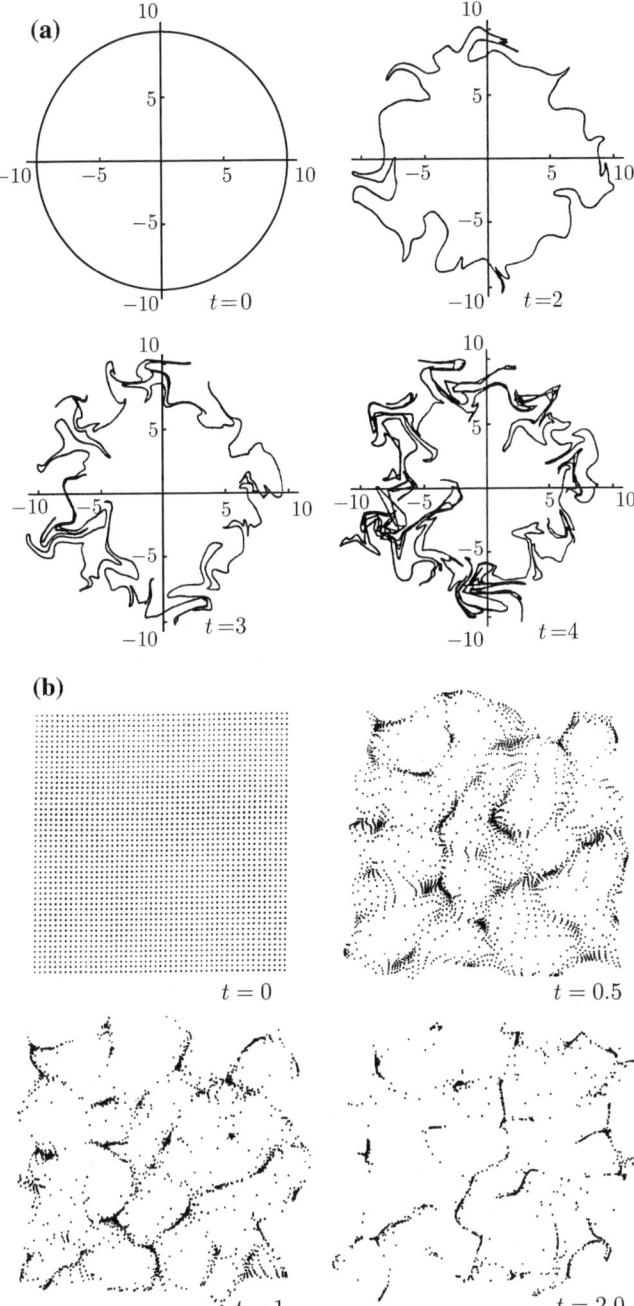

Fig. 3.1 Diffusion of a system of particles described by Eq. (3.2) numerically simulated for **a** solenoidal and **b** potential random steady velocity field **u**(**r**)

$$\frac{d}{dt}x(t) = v_x(t)\sin(2kx), \qquad x(0) = x_0,$$

$$\frac{d}{dt}\mathbf{R}(t) = \mathbf{v_R}(t)\sin(2kx), \quad \mathbf{R}(0) = \mathbf{R}_0. \tag{3.5}$$

The solution of the first equation in system (3.5) is

$$x(t) = \frac{1}{k}\arctan\left[e^{T(t)}\mathrm{tg}(kx_0)\right], \tag{3.6}$$

where

$$T(t) = 2k\int_0^t d\tau\, v_x(\tau), \tag{3.7}$$

and the solution to the second equation in (3.5) can be written in the form

$$\mathbf{R}(t|\mathbf{r}_0) = \mathbf{R}_0 + \int_0^t d\tau \frac{\sin(2kx_0)\mathbf{v_R}(\tau)}{e^{-T(\tau)}\cos^2(kx_0) + e^{T(\tau)}\sin^2(kx_0)}. \tag{3.8}$$

Consequently, if the initial particle position x_0 is such that

$$kx_0 = n\frac{\pi}{2}, \tag{3.9}$$

where $n = 0, \pm 1, \ldots$, then the particle will be the fixed particle and $\mathbf{r}(t) \equiv \mathbf{r}_0$.

Equalities (3.9) define planes in the general case and points in the one-dimensional case. They correspond to zeros of the field of velocities. Stability of these points depends on the sign of function $\mathbf{v}(t)$, and this sign changes during the evolution process. It can be expected that particles will gather around these points if $v_x(t) \neq 0$, which just corresponds to clustering of particles.

In the case of a nondivergent velocity field, $v_x(t) = 0$ and, consequently, $T(t) \equiv 0$; as a result, we have

$$x(t|x_0) \equiv x_0, \quad \mathbf{R}(t|\mathbf{r}_0) = \mathbf{R}_0 + \sin(2kx_0)\int_0^t d\tau\, \mathbf{v_R}(\tau),$$

which means that no clustering occurs.

In numerical modeling of various phenomena, a model of Gaussian vector random process $\mathbf{v}(t)$, delta-correlated in time, has been used, with the parameters

$$\langle \mathbf{v}(t) \rangle = 0 \quad \langle v_i(t)v_j(t') \rangle = 2\sigma^2\delta_{ij}\tau_0\delta(t - t'), \tag{3.10}$$

where σ^2 is the variance for each velocity component, and τ_0 its temporal correlation radius. We will adopt dimensionless variables

$$t \to k^2 \sigma^2 \tau_0 t, \quad x \to kx, \quad \langle v_i(t) v_j(t') \rangle \to 2\delta_{ij}\delta(t - t'). \tag{3.11}$$

Figure 3.2a shows a fragment of the realization of *Wiener random process* $T(t)$ obtained by numerical integration of Eq. (3.7) for a realization of random process $v_x(t)$; we used this fragment for simulating the temporal evolution of coordinates of four particles $x(t)$, $x \in (0, \pi/2)$ initially located at coordinates $x_0(i) = \dfrac{\pi}{2}\dfrac{i}{5}$ ($i = 1, 2, 3, 4$) (see Fig. 3.2b). Figure 3.2b shows that particles form a cluster in the vicinity of point $x = 0$ at the dimensionless time $t \approx 4$ (see [71]). Further, at time $t \approx 16$ the initial cluster disappears and new one appears in the vicinity of point $x = \pi/2$. At moment $t \approx 40$, the cluster appears again in the vicinity of point $x = 0$, and so on. In this process, particles in clusters remember their past history and significantly diverge during intermediate temporal segments (see Fig. 3.2c).

Thus, we see in this example that the cluster does not move from one region to another; instead, it first collapses and then a new cluster appears. Moreover, the lifetime of clusters significantly exceeds the duration of intermediate segments. It

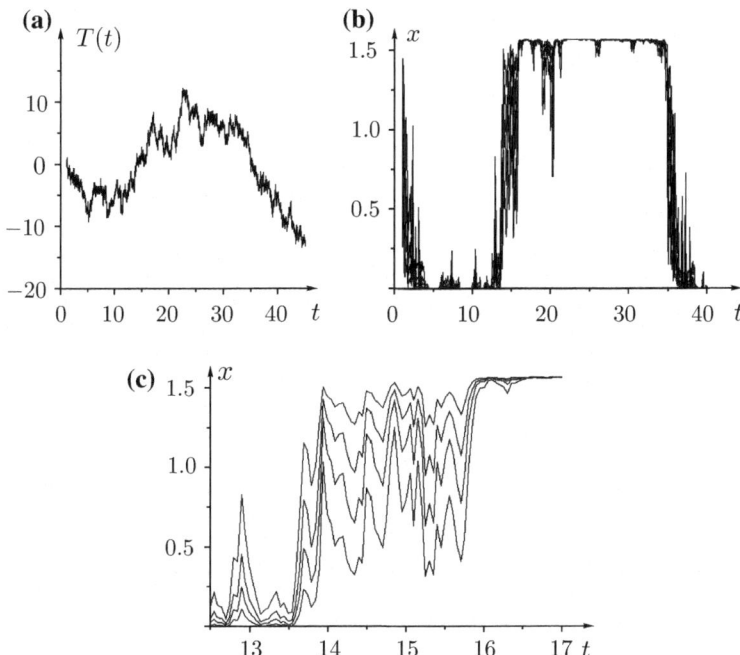

Fig. 3.2 a Segment of a realization of random process $T(t)$ obtained by numerically integrating Eq. (3.7) for a realization of random process $v_x(t)$; **b, c** x-coordinates simulated with this segment for four particles versus time

Fig. 3.3 Balloon
distribution in the
atmosphere for day 105 from
the beginning of process
simulation

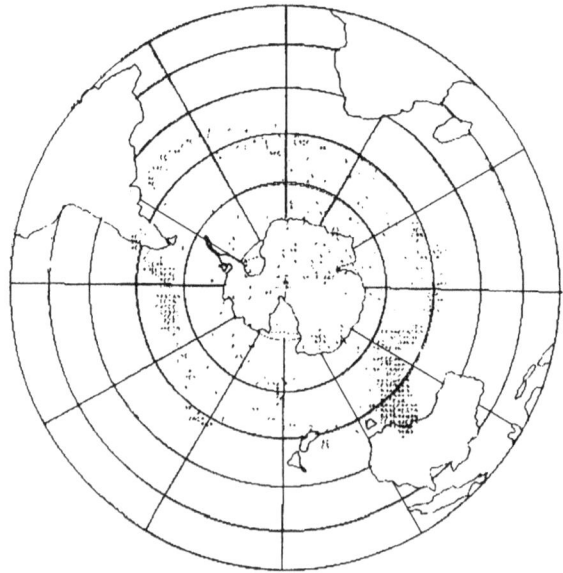

seems that this feature is characteristic of the specific model of velocity field and
follows from steadiness of points (3.9).

As regards the particle diffusion along the y-direction, no cluster occurs there.

Note that such clustering in a system of particles was found, to all appearance for
the first time as a result of simulating the so-called *Eole experiment* with the use of
the simplest equations of atmospheric dynamics.

In this global experiment, 500 constant-density balloons were launched in
Argentina in 1970–1971; these balloons traveled at a height of about 12 km and
spread along the whole of the southern hemisphere.

Figure 3.3 shows the balloon distribution over the southern hemisphere for day
105 from the beginning of this process simulation; this distribution clearly shows
that balloons are concentrated in groups, which just corresponds to clustering.

3.2 Plane Waves in Layered Media

Let the layer of inhomogeneous medium occupies the part of space $L_0 < x < L$ and
let the unit-amplitude plane wave $u_0(x) = e^{-ik(x-L)}$ is incident on this layer from
the region $x > L$ (Fig. 3.4a).

The wavefield satisfies the *Helmholtz equation*,

$$\frac{d^2}{dx^2}u(x) + k^2(x)u(x) = 0, \tag{3.12}$$

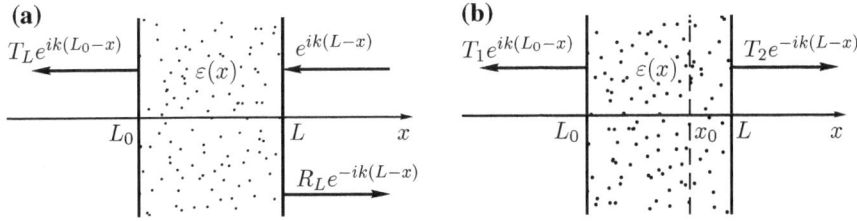

Fig. 3.4 **a** Plane wave incident on the medium layer and **b** source inside the medium layer

where

$$k^2(x) = k^2[1 + \varepsilon(x)],$$

and function $\varepsilon(x)$ describes medium inhomogeneities.

We assume that $\varepsilon(x) = 0$, i.e., $k(x) = k$ outside the layer; inside the layer, we set $\varepsilon(x) = \varepsilon_1(x) + i\gamma$, where the real part $\varepsilon_1(x)$ is responsible for wave scattering in the medium and the imaginary part $\gamma \ll 1$ describes absorption of the wave in the medium.

In region $x > L$, the wavefield has the structure

$$u(x) = e^{ik(L-x)} + R_L e^{-ik(L-x)},$$

where R_L is the complex reflection coefficient.

In region $x < L_0$, the structure of the wavefield is

$$u(x) = T_L e^{ik(L_0-x)},$$

where T_L is the complex transmission coefficient.

Boundary conditions for Eq. (3.12) are the continuity conditions for the field and the field derivative at layer boundaries; they can be written as follows

$$u(L) + \frac{i}{k}\frac{du(x)}{dx}\bigg|_{x=L} = 2, \qquad u(L_0) - \frac{i}{k}\frac{du(x)}{dx}\bigg|_{x=L_0} = 0. \qquad (3.13)$$

Thus, the wavefield in the layer of an inhomogeneous medium is described by the boundary-value problem (3.12), (3.13). Note that the problem under consideration assumes that function $\varepsilon(x)$ is discontinuous at layer boundaries. We will call the boundary-value problem (3.12), (3.13) the *unmatched boundary-value problem*. In such problems, wave scattering is caused not only by medium inhomogeneities, but also by discontinuities of function $\varepsilon(x)$ at layer boundaries.

If medium parameters (function $\varepsilon_1(x)$) are specified in the statistical form, then solving the stochastic problem (3.12), (3.13) consists in obtaining statistical characteristics of the reflection and transmission coefficients, which are related to the wavefield values at layer boundaries by the relationships

$$R_L = u(L) - 1, \quad T_L = u(L_0),$$

and the wavefield intensity

$$I(x) = |u(x)|^2$$

inside the inhomogeneous medium. Determination of these characteristics constitutes the subject of the *statistical theory of radiative transfer*.

Note that, for $x < L$, from Eq. (3.12) follows the equality

$$k\gamma I(x) = \frac{d}{dx} S(x), \tag{3.14}$$

where energy-flux density $S(x)$ is determined by the relationship

$$S(x) = \frac{i}{2k} \left[u(x) \frac{d}{dx} u^*(x) - u^*(x) \frac{d}{dx} u(x) \right].$$

By virtue of boundary conditions, we have $S(L) = 1 - |R_L|^2$ and $S(L_0) = |T_L|^2$.

As a result, integrating Eq. (3.14) over the space occupied by inhomogeneous medium, we obtain the expression

$$|R_L|^2 + |T_L|^2 + k\gamma \int_{L_0}^{L} dx I(x) = 1. \tag{3.15}$$

For non-absorptive media ($\gamma = 0$), conservation of energy-flux density is expressed by the equality

$$|R_L|^2 + |T_L|^2 = 1. \tag{3.16}$$

Consider some features characteristic of solutions to the stochastic boundary-value problem (3.12), (3.13). On the assumption that medium inhomogeneities are absent ($\varepsilon_1(x) = 0$) and absorption γ is sufficiently small, the intensity of the wavefield in the medium slowly decays with distance according to the exponential law

$$I(x) = |u(x)|^2 = e^{-k\gamma(L-x)}. \tag{3.17}$$

Figure 3.5 shows two realizations of the intensity of a wave in a sufficiently thick layer of medium. These realizations were simulated for two realizations of medium inhomogeneities. The difference between them consists in the fact that the corresponding functions $\varepsilon_1(x)$ have different signs in the middle of the layer at a distance of the wavelength. This offers a possibility of estimating the effect of a small medium mismatch on the solution of the boundary problem. Omitting the detailed description of problem parameters, we mention only that this figure clearly shows the prominent tendency of a sharp exponential decay (accompanied by significant spikes toward both higher and nearly zero-valued intensity values), which is caused by multiple

Fig. 3.5 Dynamic localization phenomenon simulated for two realizations of medium inhomogeneities

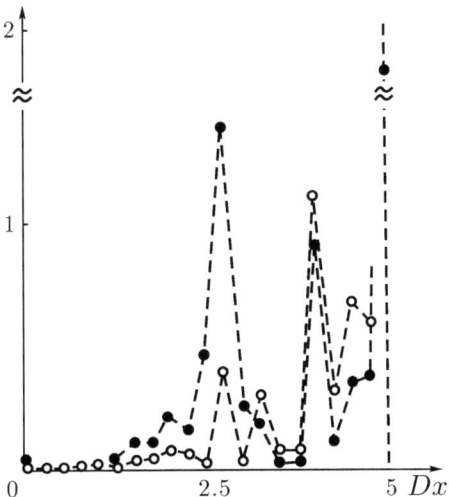

reflections of the wave in the chaotically inhomogeneous random medium (the phenomenon of *dynamic localization*). Recall that absorption is small ($\gamma \ll 1$), so that it cannot significantly affect the dynamic localization.

It is well known that the introduction of new function

$$\psi(x) = \frac{i}{k}\frac{d}{dx}\ln u(x)$$

reduces the second-order equation (3.12) to two first-order equations, and this function satisfies the closed equation following from Eq. (3.12):

$$\frac{d}{dx}\psi(x) = ik\left[\psi^2(x) - 1 - \varepsilon(x)\right], \quad \psi(L_0) = 1. \tag{3.18}$$

From the condition at boundary $x = L$ follows that

$$u(L) = \frac{2}{1 + \psi(L)}$$

and, consequently, the reflection coefficient is determined from the solution to Eq. (3.18) by the formula

$$R_L = \frac{1 - \psi(L)}{1 + \psi(L)}.$$

Introducing the function

$$R(x) = \frac{1 - \psi(x)}{1 + \psi(x)}, \quad \psi(x) = \frac{1 - R(x)}{1 + R(x)},$$

we can rewrite Eq. (3.18) in the form of the *Riccati equation*

$$\frac{d}{dx}R(x) = 2ikR(x) + \frac{i}{2k}\varepsilon(x)(1 + R(x))^2, \quad R(L_0) = 0 \quad (3.19)$$

whose solution at $x = L$ coincides with the reflection coefficient, i.e.,

$$R_L = R(L).$$

In terms of function $R(x)$, the wavefield $u(x)$ inside the medium is now expressed by the equality

$$u(x) = [1 + R(L)]\exp\left[ik\int_x^L d\xi \frac{1 - R(\xi)}{1 + R(\xi)}\right]. \quad (3.20)$$

Figure 3.6a shows the traditional procedure of solving the problem. One solves Eq. (3.19) first and then reconstructs the wavefield by the formula (3.20). This is the well known approach called the sweep method. However, it is inappropriate for analyzing statistical problems.

The *imbedding method* (see, for example, monographs [28, 29]) offers a possibility of reformulating the boundary-value problem (3.12), (3.13) to the dynamic problem with the initial values for parameter L (this parameter corresponds to the geometrical position of the layer right-hand boundary) by considering the solution to the boundary-value problem as a function of parameter L. On such reformulation, the reflection coefficient R_L satisfies the *Riccati equation*

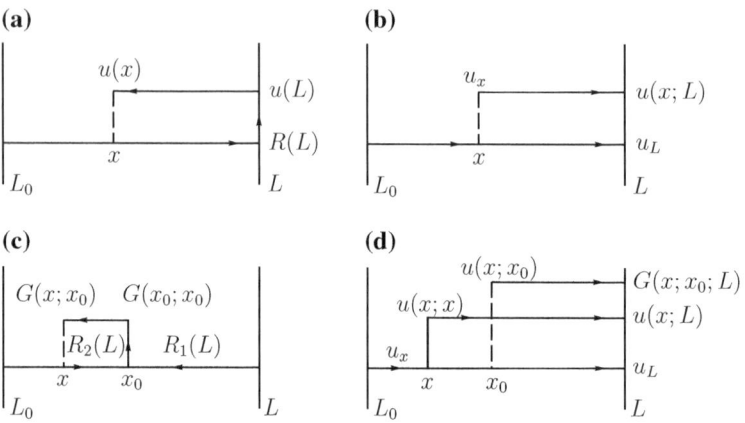

Fig. 3.6 Schematic of solving boundary problem (3.12), (3.13) by **a** the sweep method and **b** the imbedding method and boundary-value problem (3.29) by **c** the sweep method and **d** the imbedding method

$$\frac{d}{dL}R_L = 2ikR_L + \frac{ik}{2}\varepsilon(L)(1+R_L)^2, \quad R_{L_0} = 0 \tag{3.21}$$

and the wavefield in the medium layer $u(x) \equiv u(x; L)$ satisfies the linear equation

$$\frac{\partial}{\partial L}u(x; L) = iku(x; L) + \frac{ik}{2}\varepsilon(L)(1+R_L)u(x; L),$$
$$u(x; x) = 1 + R_x, \tag{3.22}$$

which can be derived, for example, by differentiating Eq. (3.20) with respect to parameter L. Figure 3.6b shows the procedure of solving the problem in this formulation. Comparing this procedure with that of the sweep method (Fig. 3.6a), we see that solving procedure has changed the direction, and namely this fact will offer a possibility of constructing the statistical description of the solution to the problem in the stochastic formulation.

From Eqs. (3.21) and (3.22) follow the equations for the squared modulus of the reflection coefficient $W_L = |R_L|^2$ and the wavefield intensity $I(x; L) = |u(x; L)|^2$

$$\frac{d}{dL}W_L = -\frac{k\gamma}{2}\left[4W_L + (R_L + R_L^*)(1+W_L)\right]$$
$$- \frac{ik}{2}\varepsilon_1(L)(R_L - R_L^*)(1-W_L), \quad W_{L_0} = 0,$$

$$\frac{\partial}{\partial L}I(x; L) = -\frac{k\gamma}{2}(2+R_L+R_L^*)I(x; L)$$
$$+ \frac{ik}{2}\varepsilon_1(L)(R_L - R_L^*)I(x; L), \quad I(x; x) = |1+R_x|^2, \tag{3.23}$$

or, after rearrangement,

$$-\frac{d}{dL}\ln(1-W_L) = -\frac{k\gamma}{2}\frac{4W_L + (R_L + R_L^*)(1+W_L)}{1-W_L}$$
$$- \frac{ik}{2}\varepsilon_1(L)(R_L - R_L^*), \tag{3.24}$$

$$\frac{\partial}{\partial L}\ln I(x; L) = -\frac{k\gamma}{2}(2+R_L+R_L^*) + \frac{ik}{2}\varepsilon_1(L)(R_L - R_L^*).$$

Excluding from Eq. (3.24) terms containing $\varepsilon_1(L)$, we obtain the equality

$$\frac{\partial}{\partial L}\ln\frac{I(x, L)}{1-W_L} = -k\gamma\frac{|1+R_L|^2}{1-W_L}.$$

Consequently, the wavefield intensity is related to the reflection coefficient by the expression

$$I(x, L) = \frac{|1 + R_x|^2 (1 - W_L)}{1 - W_x} \exp\left\{-k\gamma \int_x^L d\xi \frac{|1 + R_\xi|^2}{1 - W_\xi}\right\}. \qquad (3.25)$$

Setting $x = L_0$ in Eq. (3.25), we express the modulus of the transmission coefficient in terms of the reflection coefficient

$$|T_L|^2 = (1 - W_L) \exp\left\{-k\gamma \int_{L_0}^L d\xi \frac{|1 + R_\xi|^2}{1 - W_\xi}\right\}. \qquad (3.26)$$

In the case of non-absorptive medium, from Eq. (3.25) follows the expression

$$I(x, L) = \frac{|1 + R_x|^2 (1 - W_L)}{1 - W_x}, \qquad (3.27)$$

and one must bear in mind the relationship

$$|R_L|^2 + |T_L|^2 = 1.$$

Thus, in the case of non-absorptive medium, Eq. (3.23) can be integrated in analytic form; the resulting wavefield intensity inside the inhomogeneous layer is explicitly expressed in terms of the layer reflection coefficient.

Note that condition $W_{L_0} = 1$ will be the initial condition to Eq. (3.23) in the case of totally reflecting boundary at L_0. In this case, the wave incident on the layer of a non-absorptive medium ($\gamma = 0$) is totally reflected from the layer, i.e., $W_L = 1$.

In the general case of arbitrarily reflecting boundary L_0, the steady-state (independent of L) solution $W_L = 1$ corresponding to the total reflection of incident wave formally exists for a half-space ($L_0 \to -\infty$) filled with non-absorptive random medium, too. As will be shown below, this solution, is actually realized in the statistical problem with probability equal to one.

If, in contrast to the above problem, we assume that function $k(x)$ is continuous at boundary $x = L$, i.e., if we assume that the wave number in the free half-space $x > L$ is equal to $k(L)$, then boundary conditions (3.13) of problem (3.12) will be replaced with the conditions

$$u(L) + \frac{i}{k(L)} \frac{du(x)}{dx}\bigg|_{x=L} = 2, \quad u(L_0) - \frac{i}{k(L_0)} \frac{du(x)}{dx}\bigg|_{x=L_0} = 0. \qquad (3.28)$$

We will call the boundary-value problem (3.12), (3.28) the *matched boundary-value problem*.

The field of a point source located in the layer of random medium is described by the similar boundary-value problem for Green's function of the Helmholtz equation:

$$\frac{d^2}{dx^2}G(x; x_0) + k^2[1 + \varepsilon(x)]G(x; x_0) = 2ik\delta(x - x_0),$$

$$G(L; x_0) + \frac{i}{k}\frac{dG(x; x_0)}{dx}\bigg|_{x=L} = 0, \quad G(L_0; x_0) - \frac{i}{k}\frac{dG(x; x_0)}{dx}\bigg|_{x=L_0} = 0.$$

$$\text{(3.29)}$$

Outside the layer, the solution has here the form of outgoing waves (Fig. 3.4b)

$$G(x; x_0) = T_1 e^{ik(x-L)} \quad (x \ge L), \qquad G(x; x_0) = T_2 e^{-ik(x-L_0)} \quad (x \le L_0).$$

Note that, for the source located at the layer boundary $x_0 = L$, this problem coincides with the boundary-value problem (3.12), (3.13) on the wave incident on the layer, which yields

$$G(x; L) = u(x; L).$$

The solution of boundary-value problem (3.29) has the structure

$$G(x; x_0) = G(x_0; x_0) \begin{cases} \exp\left[ik\int_x^{x_0}\psi_1(\xi)\,d\xi\right], & x_0 \ge x, \\[2em] \exp\left[ik\int_{x_0}^x\psi_2(\xi)\,d\xi\right], & x_0 \le x, \end{cases} \qquad \text{(3.30)}$$

where the field at the source location, by virtue of the derivative gap condition

$$\frac{dG(x; x_0)}{dx}\bigg|_{x=x_0+0} - \frac{dG(x; x_0)}{dx}\bigg|_{x=x_0-0} = 2ik,$$

is determined by the formula

$$G(x_0; x_0) = \frac{2}{\psi_1(x_0) + \psi_2(x_0)},$$

and functions $\psi_i(x)$ satisfy the Riccati equations

$$\frac{d}{dx}\psi_1 = ik\left[\psi_1^2 - 1 - \varepsilon(x)\right], \quad \psi_1(L_0) = 1,$$

$$\frac{d}{dx}\psi_2 = -ik\left[\psi_2^2 - 1 - \varepsilon(x)\right], \quad \psi_2(L) = 1.$$

$$\text{(3.31)}$$

Figure 3.6c shows the procedure of solving this problem by the sweep method. One solves two Eq. (3.31) first and then reconstructs the wavefield using Eq. (3.30). Introduce new functions $R_i(x)$ related to functions $\psi_i(x)$ by the formula

$$\psi_i(x) = \frac{1 - R_i(x)}{1 + R_i(x)}, \quad i = 1, 2.$$

With these functions, the wavefield in region $x < x_0$ can be written in the form

$$G(x; x_0) = \frac{[1 + R_1(x_0)][1 + R_2(x_0)]}{1 - R_1(x_0)R_2(x_0)} \exp\left[ik \int_x^{x_0} d\xi \frac{1 - R_1(\xi)}{1 + R_1(\xi)}\right], \quad (3.32)$$

where function $R_1(x)$ satisfies the Riccati equation (3.19).

For $x_0 = L$, expression (3.32) becomes

$$G(x; L) = u(x; L) = [1 + R_1(L)]\exp\left[ik \int_x^L d\xi \frac{1 - R_1(\xi)}{1 + R_1(\xi)}\right], \quad (3.33)$$

so that parameter $R_1(L) = R_L$ is the reflection coefficient of the plane wave incident on the layer from region $x > L$. In a similar way, quantity $R_2(x_0)$ is the reflection coefficient of the wave incident on the medium layer (x_0, L) from the homogeneous half-space $x < x_0$ (i.e., from region with $\varepsilon = 0$).

Using Eq. (3.33), we can rewrite Eq. (3.32) in the form

$$G(x; x_0) = \frac{1 + R_2(x_0)}{1 - R_1(x_0)R_2(x_0)}u(x; x_0), \quad x \le x_0,$$

where $u(x; x_0)$ is the wavefield inside the inhomogeneous layer (L_0, x_0) in the case of the incident wave coming from the free half-space $x > x_0$.

Thus, for $x < x_0$, the field of the point source is proportional to the wavefield generated by the plane wave incident on layer (L_0, x_0) from the free half-space $x > x_0$. The layer segment (x_0, L) affects only parameter $R_2(x_0)$.

Problems with perfectly reflecting boundaries at which either $G(x; x_0)$ or $\frac{d}{dx}G(x; x_0)$ vanishes are of great interest for applications. Indeed, in the latter case, we have $R_2(x_0) = 1$ for the source located at this boundary; consequently,

$$G_{\text{ref}}(x; x_0) = \frac{2}{1 - R_1(x_0)} \exp\left[ik \int_x^{x_0} d\xi \frac{1 - R_1(\xi)}{1 + R_1(\xi)}\right], \quad x \le x_0. \quad (3.34)$$

In addition, the expression for wavefield intensity $I(x; x_0) = |G(x; x_0)|^2$ follows from Eq. (3.29) for $x < x_0$

$$k\gamma I(x; x_0) = \frac{d}{dx}S(x; x_0), \quad (3.35)$$

where energy flux density $S(x; x_0)$ is given by the expression

$$S(x; x_0) = \frac{i}{2k} \left[G(x; x_0) \frac{d}{dx} G^*(x; x_0) - G^*(x; x_0) \frac{d}{dx} G(x; x_0) \right].$$

Using Eq. (3.32), we can represent $S(x; x_0)$ in the form $(x \leq x_0)$

$$S(x; x_0) = S(x_0; x_0) \exp \left[-k\gamma \int\limits_x^{x_0} d\xi \frac{|1 + R_1(\xi)|^2}{1 - |R_1(\xi)|^2} \right],$$

where the energy flux density at the point of source location is

$$S(x_0; x_0) = \frac{\left[1 - |R_1(x_0)|^2 \right] |1 + R_2(x_0)|^2}{|1 - R_1(x_0) R_2(x_0)|^2}. \tag{3.36}$$

Below, our concern will be with statistical problems on waves incident on random half-space $(L_0 \to -\infty)$ and source-generated waves in infinite space $(L_0 \to -\infty, L \to \infty)$ for sufficiently small absorption $(\gamma \to 0)$. One can see from Eq. (3.35) that these limit processes are not commutable in the general case. Indeed, if $\gamma = 0$, then energy flux density $S(x; x_0)$ is conserved in the whole half-space $x < x_0$. However, integrating Eq. (3.35) over half-space $x < x_0$ in the case of small but finite absorption, we obtain the restriction on the energy confined in this half-space

$$k\gamma \int\limits_{-\infty}^{x_0} dx\, I(x; x_0) = S(x_0; x_0) = \frac{\left[1 - |R_1(x_0)|^2 \right] |1 + R_2(x_0)|^2}{|1 - R_1(x_0) R_2(x_0)|^2}. \tag{3.37}$$

Note that, considering the wavefield as a function of parameter L (i.e., setting $G(x; x_0) \equiv G(x; x_0; L)$), we can use the imbedding method to obtain the following system of equations with initial values:

$$\frac{\partial}{\partial L} G(x; x_0; L) = i\frac{k}{2} \varepsilon(L) u(x_0; L) u(x; L),$$

$$G(x; x_0; L)|_{L=\max(x, x_0)} = \begin{cases} u(x; x_0), & x \geq x_0 \\ u(x_0; x), & x \leq x_0 \end{cases},$$

$$\frac{\partial}{\partial L} u(x; L) = ik \{ 1 + \varepsilon(L) u(L; L) \} u(x; L), \tag{3.38}$$

$$u(x; L)|_{L=x} = u(x; x),$$

$$\frac{d}{dL} u(L; L) = 2ik[u(L; L) - 1] + i\frac{k}{2} \varepsilon(L) u^2(L; L),$$

$$u(L_0; L_0) = 1.$$

Here, two last equations describe the wavefield appearing in the problem on the wave incident on the medium layer. Figure 3.6d shows the procedure of solving this problem.

Statistical description of these boundary-value problems will be discussed in detail in Chap. 7. The *Anderson dynamic localization* is also known for eigenfunctions of the one-dimensional stationary Schrödinger (Gelmholtz) equation with a random potential $\varepsilon(x)$ [46, 47].

3.3 Partial Differential Equations

As for random fields, we may introduce a generalization of lognormal random process (2.2), extending it to a lognormal random field according to the formula

$$f(\mathbf{r}, t; \alpha) = f_0(\mathbf{r}) \exp\left\{-\alpha t + \int_0^t d\tau z(\mathbf{r}, \tau)\right\}, \tag{3.39}$$

where $z(\mathbf{r}, t)$ is the Gaussian random field delta-correlated in time with a zero mean and the correlation function

$$B_z(\mathbf{r} - \mathbf{r}', t - t') = \langle z(\mathbf{r}, t) z(\mathbf{r}', t') \rangle = 2D(\mathbf{r} - \mathbf{r}')\delta(t - t'). \tag{3.40}$$

This field satisfies the first-order differential equation

$$\frac{d}{dt} f(\mathbf{r}, t; \alpha) = \{-\alpha + z(\mathbf{r}, t)\} f(\mathbf{r}, t; \alpha), \quad f(\mathbf{r}, 0; \alpha) = f_0(\mathbf{r}), \tag{3.41}$$

which parametrically depends on the position of point $\mathbf{r}$ in space.

Notice that the studies [37, 38] on intermittency mentioned above considered the equation

$$\frac{d}{dt} f(\mathbf{r}, t) = z(\mathbf{r}, t) f(\mathbf{r}, t) + \mu_f \Delta f(\mathbf{r}, t) \tag{3.42}$$

as their model problem, where μ_f is the dynamic diffusion coefficient for the field $f(\mathbf{r}, t)$. This equation with random breeding and diffusion is typical of problems in biology and kinetics of chemical and nuclear reactions (see, for example, [50]).

If $f_0(\mathbf{r}) = 1$, all one-point statistical characteristics of this field are independent of $\mathbf{r}$.

At the initial stage of diffusion, the solution to problem (3.39) is given by function (3.39) with $\alpha = 0$ and $f_0(\mathbf{r}) = 1$:

$$f(\mathbf{r}, t) = \exp\left\{\int_0^t d\tau z(\mathbf{r}, \tau)\right\}, \tag{3.43}$$

which is statistically equivalent to the random process $y(t; 0)$ (2.2) for one-point statistical characteristics. As will be shown further, there is no structure formation in this case. But the general feature of intermittency remains.

It should be noted that adding the 'destruction' effect to Eq. (3.42) by introducing the term $\{-\alpha f(\mathbf{r}, t; \alpha)\}$ for $\alpha > 0$ leads to the equation

$$\frac{d}{dt} f(\mathbf{r}, t; \alpha) = \{-\alpha + z(\mathbf{r}, t)\} f(\mathbf{r}, t; \alpha) + \mu_f \Delta f(\mathbf{r}, t; \alpha), \qquad (3.44)$$

the solution of which at the initial stage is already described by formula (3.39). In this case, as shown further, stochastic structure formation in the form of clustering becomes possible.

Randomness in medium parameters gives rise to a stochastic behavior of physical fields. Individual samples of scalar two-dimensional fields $f(\mathbf{R}, t)$ (3.43), where $\mathbf{R} = \{x, y\}$, say, recall a rough mountainous terrain with randomly scattered peaks, troughs, ridges and saddles. Figure 3.7 illustrates examples of two numerically simulated realizations of two random fields with different statistical structures.

Clustering in random physical fields arises first of all in problems of *turbulent transport* in a random velocity field $\mathbf{u}(\mathbf{r}, t)$. In particular, *clustering* may occur for both *a passive scalar tracer* (the field of density) [28, 29, 49], and *a vector tracer*

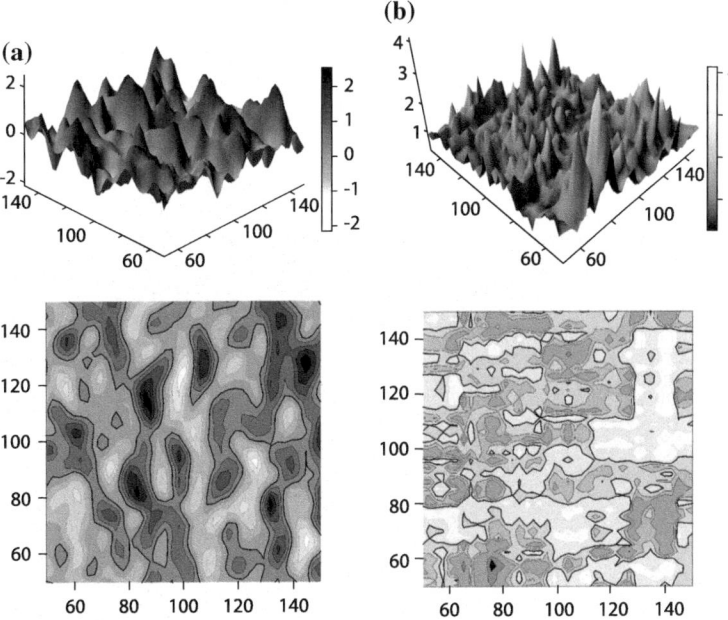

Fig. 3.7 Realizations of the fields governed by **a** Gaussian and **b** lognormal distributions and the corresponding topographic level lines. The *bold curves* in the bottom patterns show *level lines* corresponding to levels 0 (**a**) and 1 (**b**)

(magnetic field energy) in the framework of *kinematic approximation* [28, 29, 51]. The input stochastic equations in these cases are the continuity equation for the tracer density field $\rho(\mathbf{r}, t)$, viz.

$$\left(\frac{\partial}{\partial t} + \frac{\partial}{\partial \mathbf{r}}\mathbf{u}(\mathbf{r}, t)\right)\rho(\mathbf{r}, t) = \mu_\rho\Delta\rho(\mathbf{r}, t), \quad \rho(\mathbf{r}, 0) = \rho_0(\mathbf{r}), \tag{3.45}$$

and the induction equation for solenoidal magnetic field $\mathbf{H}(\mathbf{r}, t)$ [52]:

$$\left(\frac{\partial}{\partial t} + \frac{\partial}{\partial \mathbf{r}}\mathbf{u}(\mathbf{r}, t)\right)\mathbf{H}(\mathbf{r}, t) = \left(\mathbf{H}(\mathbf{r}, t) \cdot \frac{\partial}{\partial \mathbf{r}}\right)\mathbf{u}(\mathbf{r}, t) + \mu_H\Delta\mathbf{H}(\mathbf{r}, t),$$
$$\mathbf{H}(\mathbf{r}, 0) = \mathbf{H}_0(\mathbf{r}), \tag{3.46}$$

where μ_ρ and μ_H are the dynamic diffusion coefficients for the density field and the magnetic field, respectively. Here, $\mathbf{u}(\mathbf{r}, t)$ is the field of turbulent velocities with given statistical properties, which is assumed to be homogeneous and isotropic in space, and stationary in time.

We stress that in the analysis of these and similar equations of mathematical physics, considered further in this work, we will not be interested in their direct solutions or physical mechanisms given birth to one physical phenomenon or another. Our goal is to learn whether the input equations on their own contain information on the possibility (or impossibility) of stochastic structure formation in random media with a unit probability, i.e., for almost all realizations of their solutions.

It should be noted that a scalar density field always experiences clustering in a compressible fluid flow. Figure 3.8 displays the pattern of cluster structure of the Universe, taken from the Internet, which in all probability is directly related to cosmic matter clustering in random velocity fields. This question is discussed in Sect. 9.1.

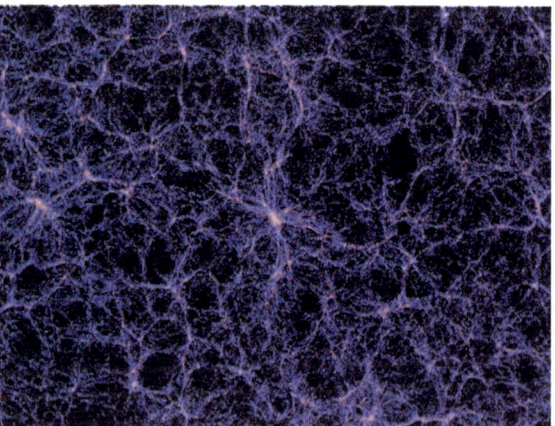

Fig. 3.8 Cluster structure of the Universe

The formation of stochastic structures may also take the form of a caustic structure of the wave field intensity in problems involving waves propagating in randomly inhomogeneous media in the framework of the Leontovich complex-valued stochastic parabolic equation (see, for example, [28, 54]):

$$\frac{\partial}{\partial x}u(x, \mathbf{R}) = \frac{i}{2k}\Delta_\mathbf{R}u(x, \mathbf{R}) + \frac{ik}{2}\varepsilon(x, \mathbf{R})u(x, \mathbf{R}), \quad u(x, \mathbf{R}) = u_0(\mathbf{R}), \quad (3.47)$$

where i is the imaginary unit, x is the coordinate in the direction of wave propagation, $\mathbf{R}$ is the coordinate in the transverse plane, and, $\varepsilon(x, \mathbf{R})$ is the deviation of dielectric constant from unity.

We mention that this same equation becomes a nonstationary Schrödinger equation with the random potential $\varepsilon(x, \mathbf{R})$ on the replacement of x with time t.

If we introduce the amplitude and phase of a wave field according to the formula

$$u(x, \mathbf{R}) = A(x, \mathbf{R})\exp\{i\,S(x, \mathbf{R})\},$$

the equation for the wave field intensity $I(x, \mathbf{R}) = |u(x, \mathbf{R})|^2$ takes the following form:

$$\frac{\partial}{\partial x}I(x, \mathbf{R}) + \frac{1}{k}\nabla_\mathbf{R}\{\nabla_\mathbf{R}S(x, \mathbf{R})I(x, \mathbf{R})\} = 0, \quad I(0, \mathbf{R}) = I_0(\mathbf{R}). \quad (3.48)$$

Equation (3.48) coincides in form with the continuity equation (3.45) for the tracer density field in a random potential flow in the absence of dynamic diffusivity and, accordingly, the wave field intensity should undergo clustering. This problem is discussed in more detail in Sect. 8.2.1.

It should also be emphasized that a nonlinear generalization of Eq. (3.47), which corresponds to a monochromatic nonlinear problem on wave self-action in randomly inhomogeneous media, is described by the Leontovich complex-valued nonlinear parabolic equation (the *nonlinear Schrödinger equation*)

$$\frac{\partial}{\partial x}u(x, \mathbf{R}) = \frac{i}{2k}\Delta_\mathbf{R}u(x, \mathbf{R}) + \frac{ik}{2}\varepsilon(x, \mathbf{R}; I(x, \mathbf{R}))u(x, \mathbf{R}), \quad u(0, \mathbf{R}) = u_0(\mathbf{R}).$$

$$(3.49)$$

For Eq. (3.49), the wave field intensity $I(x, \mathbf{R})$ is also described by Eq. (3.48) (but, clearly, with another phase function $S(x, \mathbf{R})$), so that the intensity should experience clustering, too.

In particular, since Eq. (3.48) does not depend on the form of function $\varepsilon(x, \mathbf{R})$, then even if $\varepsilon(x, \mathbf{R}) = 0$ for the initial condition $u_0(\mathbf{R})$, the caustic structure formation, as is well known, takes place, which is regularly observed in swimming pools or in shallow water. In this case, Eqs. (3.47) and (3.49) take the form

$$\frac{\partial}{\partial x}u(x, \mathbf{R}) = \frac{i}{2k}\Delta_\mathbf{R}u(x, \mathbf{R}), \quad u(0, \mathbf{R}) = u_0(\mathbf{R}).$$

Fig. 3.9 Caustics in a swimming pool, and in shallow water

The solution to the last equation is the function

$$u(x, \mathbf{R}) = \exp\left\{\frac{ix}{2k}\Delta_{\mathbf{R}}\right\} u_0(\mathbf{R}) = \frac{k}{2\pi i x}\int d\mathbf{R}' \exp\left\{\frac{ik}{2x}(\mathbf{R} - \mathbf{R}')^2\right\} u_0(\mathbf{R}')$$

(3.50)

and for a plane incident wave the initial condition is $|u_0(\mathbf{R})| = 1$, i.e., the condition $u_0(\mathbf{R}) = e^{iS_0(\mathbf{R})}$, where $S_0(\mathbf{R})$ is the field of the random initial phase. In this case, spatial fluctuations in the initial distribution of the wave phase transform into a caustic structure in the wave field intensity. The case is known as the *random phase screen*. Examples of such clustering are given in Fig. 3.9.

dynamic systems (3.45)–(3.49) are conservative and preserve integral characteristics such as the total tracer mass $M = \int d\mathbf{r}\rho(\mathbf{r}, t)$, the magnetic field flux $\int d\mathbf{r}\, \mathbf{H}(\mathbf{r}, t)$ and wave field power $I = \int d\mathbf{R} I(x, \mathbf{R})$.

For homogeneous initial conditions $\rho_0(\mathbf{r}) = \rho_0$, $\mathbf{H}_0(\mathbf{r}) = \mathbf{H}_0$, and $u_0(\mathbf{R}) = u_0$, and for random parameters that are statistically homogeneous in space, the corollary of the conservative character of dynamic systems (3.45)–(3.49) is the equalities

$$\langle\rho(\mathbf{r}, t)\rangle = \rho_0, \quad \langle\mathbf{H}(\mathbf{r}, t)\rangle = \mathbf{H}_0, \quad \langle I(x, \mathbf{R})\rangle = I_0 = |u_0|^2.$$

A peculiarity of Eqs. (3.45) and (3.46) is the parametric excitation of both the density field $\rho(\mathbf{r}, t)$ (in a compressible fluid flow) and the magnetic field energy $E(\mathbf{r}, t) = \mathbf{H}^2(\mathbf{r}, t)$ (for a turbulent flow of fluid) with time in *each realization*, which has come to be known as the *stochastic dynamo* (see, for example, [52]).

At the initial stages of dynamic system evolution such parametric excitation is accompanied by an increase with time of all traditional statistical characteristics of problem solutions, such as the moment function of the density field $\langle\rho^n(\mathbf{r}, t)\rangle$ and magnetic field energy $\langle E^n(\mathbf{r}, t)\rangle$, as well as their correlation functions of arbitrary order. As the distance is increased, the moments of radiation power $\langle I^n(x, \mathbf{R})\rangle$ grow in random media as well.

The dynamic diffusion effects for the density and magnetic field are insignificant through the early phases of their evolution and, neglecting them, we arrive at the first-order partial differential equations

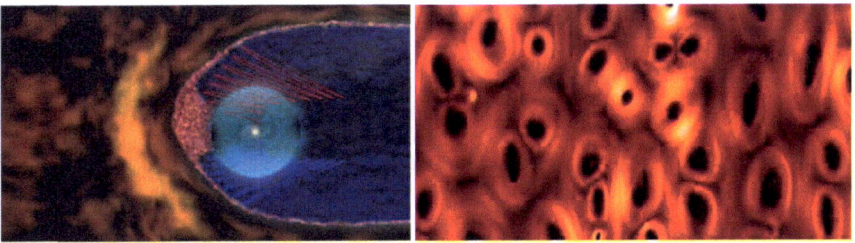

Fig. 3.10 Magnetic field configuration on the boundary of a heliosphere as it looks, in all probability, in reality. A conditional interpretation (*left*), and a reconstruction of a magnetic bubble system (*right*)

$$\left(\frac{\partial}{\partial t} + \frac{\partial}{\partial \mathbf{r}}\mathbf{u}(\mathbf{r}, t)\right)\rho(\mathbf{r}, t) = 0, \quad \rho(\mathbf{r}, 0) = \rho_0(\mathbf{r}), \tag{3.51}$$

$$\left(\frac{\partial}{\partial t} + \frac{\partial}{\partial \mathbf{r}}\mathbf{u}(\mathbf{r}, t)\right)\mathbf{H}(\mathbf{r}, t) = \left(\mathbf{H}(\mathbf{r}, t) \cdot \frac{\partial}{\partial \mathbf{r}}\right)\mathbf{u}(\mathbf{r}, t), \quad \mathbf{H}(\mathbf{r}, 0) = \mathbf{H}_0(\mathbf{r}). \tag{3.52}$$

However, namely at small times *spatial structures* may develop in individual realizations of respective fields!

As an illustration of structure formation in a magnetic field, we present Fig. 3.10 and an excerpt found on the web [53]:

"*What does puzzle astrophysicists so strongly?*

Contrary to hypotheses formed for 50 years, at the boundary of a planetary system observers encountered a boiling foam of locally magnetized areas each of hundreds of millions kilometers in extent, which form a nonstationary cellular structure in which magnetic field lines are permanently breaking and recombining to form new areas—*magnetic bubbles.*"

Questions touching stochastic structure formation for the magnetic field energy are considered in Sect. 9.2.2.

3.4 Model of a Stochastic Velocity Field Allowing Analytical Solutions to Transport Problems

Let us now consider a simple model velocity field of the form (3.3), used in Sect. 3.1.1 $\mathbf{u}(\mathbf{r}, t) = \mathbf{v}(t)f(kx)$, where $\mathbf{v}(t)$ is a is a Gaussian vector random *white noise* process with parameters (3.10), and $f(kx)$ is a periodic function. Notice that the velocity field model of the form (3.4) $\mathbf{u}(\mathbf{r}, t) = \mathbf{v}(t)\sin 2(kx)$, allowed the derivation of an analytical solution of continuity equation (3.51) in scalar density field $\rho(\mathbf{r}, t)$, as well as a solution of Eq. (3.52) in vector magnetic field and, consequently, gave the possibility of tracing the onset and evolution of cluster formation in these fields in individual realizations of the random velocity field.

Dimensionless variables (3.11) were used in the numerical simulations of different problems.

3.4.1 Model of Passive Tracer Diffusion

The solution of Eq. (3.51) for the density field in the case considered has the form [71]

$$\rho(x, t)/\rho_0 = \frac{1}{e^{T(t)} \cos^2(kx) + e^{-T(t)} \sin^2(kx)}, \tag{3.53}$$

where $T(t) = 2k \int_0^t d\tau \, v_x(\tau)$ is the Wiener random process.

From expression (3.53) it can be seen that the density field is small everywhere except for the vicinities of points $kx = n\frac{\pi}{2}$, where $\rho(x, t)/\rho_0 = e^{\pm T(t)}$, i.e., the field is rather strong close to these points, granted the sign of the random factor $T(t)$.

Thus, in the problem considered, the cluster structure of the density field in the Euler description forms in the vicinity of points

$$kx = n\frac{\pi}{2} \quad (n = 0, \pm 1, \pm 2, \ldots).$$

The results of simulations of space-time evolution experienced by a realization of the Euler density field $1 + \rho(x, t)/\rho_0$ (unity is added to eliminate difficulties in logarithmic representation for densities approaching zero) are presented in Fig. 3.11 in dimensionless variables (3.11). This figure explicitly illustrates a gradual flow of the density field to the vicinities of points $x \approx 0$ and $x \approx \pi/2$, i.e., the formation of clusters at locations where the relative value of the density is large, whereas it is close to zero in the remaining space. Notice that at time instants t, such that $T(t) = 0$, the realization of the density field passes through the initial homogeneous state.

3.4.2 Turbulent Dynamo Model

For induction equation (3.52) and model (3.3), (3.4) considered here, the x-component of the magnetic field is preserved, i.e., $H_x(\mathbf{r}, t) = H_{x0}$, and the transverse magnetic field component $\mathbf{H}_\perp(x, t)$ satisfies the equation

$$\left(\frac{\partial}{\partial t} + v_x(t) \frac{\partial}{\partial x} f(x) \right) \mathbf{H}_\perp(x, t) = \mathbf{v}_\perp(t) \frac{\partial f(x)}{\partial x} H_{x0} \tag{3.54}$$

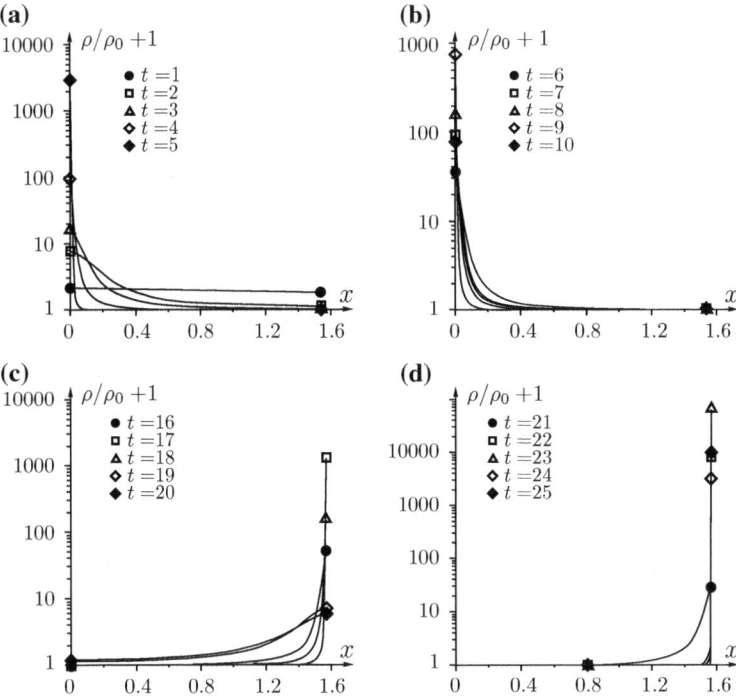

Fig. 3.11 Spatio-temporal evolution of the Eulerian density field, described by formula (3.53)

with the initial condition $\mathbf{H}_\perp(x, 0) = \mathbf{H}_{\perp 0}$; its solution can be written in a statistically equivalent form (at $\mathbf{H}_{\perp 0} = 0$) as

$$\mathbf{H}_\perp(x, t) = 2kH_{x0} \int\limits_0^t d\tau \frac{\left[e^{T(\tau)}\cos^2(kx) - e^{-T(\tau)}\sin^2(kx)\right]}{\left[e^{T(\tau)}\cos^2(kx) + e^{-T(\tau)}\sin^2(kx)\right]^2} v_x(\tau)\mathbf{v}_\perp(\tau). \quad (3.55)$$

The expression on the right-hand side of formula (3.55) describes the generation of magnetic field $\mathbf{H}_\perp(x, t)$ in the transverse plane (y, z) due to the presence of the initial field H_{x0}. And at $\mathbf{H}_{\perp 0} = 0$ field $\mathbf{H}_\perp(x, t)$, proportional to the square of the random velocity field, defines the situation. The structure of field $\mathbf{H}_\perp(x, t)$ similarly to that of density field, also experiences clustering, which is confirmed by the results of numerical simulations (see Refs. [51, 72] and the monographs [28, 29]), presented in dimensionless variables (3.11) in Fig. 3.12a, which plots the cluster share of the generated magnetic field energy with respect to the total energy in the layer at a given time instant, and in Fig. 3.12b, where we see the dynamics of how perturbations in magnetic energy flow from one domain boundary to the other one.

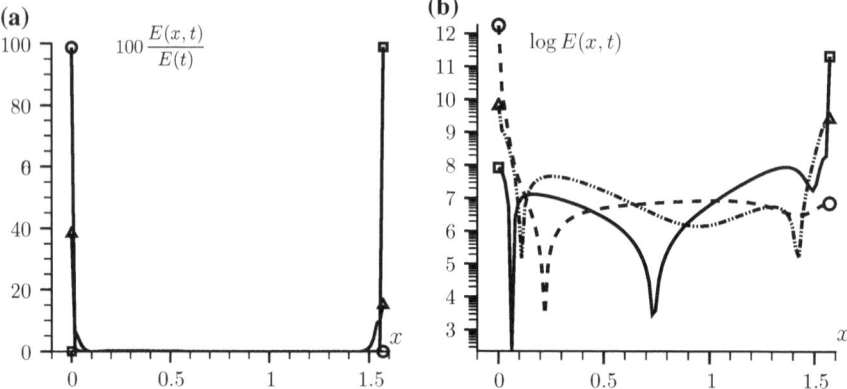

Fig. 3.12 Dynamics of cluster disappearance at the point 0 and cluster emergence at the point $\pi/2$. The *circle* stands for the time instant $t = 10.4$, *triangle* for $t = 10.8$, and *square* for $t = 11.8$

Let us point out a specific feature of Eq. (3.54). In this case, the parametric excitation of the magnetic field is accompanied by the *Gaussian generation* of the field itself.

Chapter 4
Statistical Characteristics of a Random Velocity Field u(r, t)

The random velocity field $\mathbf{u}(\mathbf{r}, t)$ will be considered Gaussian, statistically homogeneous and isotropic in space, and stationary in time, with the respective correlation and spectral functions

$$B_{ij}(\mathbf{r} - \mathbf{r}', t - t') = \langle u_i(\mathbf{r}, t) u_j(\mathbf{r}', t') \rangle = \int d\mathbf{k} E_{ij}(\mathbf{k}, t - t') e^{i\mathbf{k}(\mathbf{r} - \mathbf{r}')},$$

$$E_{ij}(\mathbf{k}, t) = \frac{1}{(2\pi)^3} \int d\mathbf{r} B_{ij}(\mathbf{r}, t) e^{-i\mathbf{k}\mathbf{r}}. \tag{4.1}$$

In the general case of an arbitrary random velocity field $\mathbf{u}(\mathbf{r}, t)$ the spectral function $E_{ij}(\mathbf{k}, t)$ has the form

$$E_{ij}(\mathbf{k}, t) = E^s(k, t) \left(\delta_{ij} - \frac{k_i k_j}{k^2} \right) + \frac{k_i k_j}{k^2} E^P(k, t), \tag{4.2}$$

where $E^s(k, t)$ and $E^P(k, t)$ are, respectively, the solenoidal and potential components of the velocity field spectral function. In this case

$$B_{ij}(\mathbf{0}, 0) = \langle u_i(\mathbf{r}, t) u_j(\mathbf{r}, t) \rangle = \frac{1}{d} \sigma_{\mathbf{u}}^2 \delta_{ij},$$

where d is the space dimension, and the variance of the velocity field takes the form

$$\sigma_{\mathbf{u}}^2 = \langle \mathbf{u}(\mathbf{r}, t)^2 \rangle = B_{ii}(\mathbf{0}, 0)$$
$$= \int d\mathbf{k} E_{ii}(\mathbf{k}, 0) = \int d\mathbf{k} \left[(d - 1) E^s(k, 0) + E^P(k, 0) \right], \tag{4.3}$$

with implied summation taken over twice repeating indices.

© Springer International Publishing AG 2017
V.I. Klyatskin, *Fundamentals of Stochastic Nature Sciences*,
Understanding Complex Systems, DOI 10.1007/978-3-319-56922-2_4

Let us introduce the function

$$B_{ij}(\mathbf{r}) = \int_0^\infty d\tau\, B_{ij}(\mathbf{r}, \tau), \tag{4.4}$$

important for the further statistical analysis, which defines all statistical characteristics of the problem solution in the diffusion approximation (see, for example, monographs [28, 29]). The value $B_{ii}(\mathbf{0}) = D_0$ is a constant, where

$$D_0 = \sigma_{\mathbf{u}}^2 \tau_0 = \int_0^\infty d\tau\, B_{ii}(\mathbf{0}, \tau) = \int_0^\infty d\tau \int d\mathbf{k}\, E_{ii}(\mathbf{k}, \tau) = \sigma_s^2 \tau_s + \sigma_P^2 \tau_P. \tag{4.5}$$

Parameter τ_0 is the time correlation radius of a random velocity field, and σ_s^2 and σ_P^2 are, respectively, the variances of the solenoidal and potential components of the velocity field, and τ_s and τ_P are their time correlation radii.

Later on, in a statistical analysis of the problem, we will need the second spatial derivatives of the correlation function of a random velocity field $\mathbf{u}(\mathbf{r}, t)$ at the zero argument. By virtue of the assumptions about the characteristics of random field $\mathbf{u}(\mathbf{r}, t)$ the following tensor equality holds for these derivatives (see, for example, monographs [28, 29]):

$$-\frac{\partial^2 B_{ij}(0)}{\partial r_k \partial r_l} = \frac{D^s}{d(d+2)}\left[(d+1)\delta_{kl}\delta_{ij} - \delta_{ki}\delta_{lj} - \delta_{kj}\delta_{li}\right]$$

$$+ \frac{D^P}{d(d+2)}\left[\delta_{kl}\delta_{ij} + \delta_{ki}\delta_{lj} + \delta_{kj}\delta_{li}\right], \tag{4.6}$$

where in the three-dimensional case the parameters D^s and D^P have the form

$$D^s = \int d\mathbf{k}\, k^2 E^s(k) = 4\pi \int_0^\infty dk\, k^4 E^s(k) = \frac{1}{2} \int_0^\infty d\tau\, \langle \omega(\mathbf{r}, t+\tau)\omega(\mathbf{r}, t) \rangle,$$

$$D^P = \int d\mathbf{k}\, k^2 E^P(k) = 4\pi \int_0^\infty dk\, k^4 E^P(k) = \int_0^\infty d\tau\, \left\langle \frac{\partial \mathbf{u}(\mathbf{r}, t+\tau)}{\partial \mathbf{r}} \frac{\partial \mathbf{u}(\mathbf{r}, t)}{\partial \mathbf{r}} \right\rangle.$$

$$\tag{4.7}$$

Here, $\omega(\mathbf{r}, t) = \mathrm{curl}\, \mathbf{u}(\mathbf{r}, t)$ is the vorticity, $\partial \mathbf{u}(\mathbf{r}, t)/\partial \mathbf{r}$ is the divergence of the velocity field, and

$$E^s(k) = \int_0^\infty d\tau\, E^s(k, \tau), \quad E^P(k) = \int_0^\infty d\tau\, E^P(k, \tau).$$

The coefficients D^s and D^P, defined by relationships (4.7), can be written out through the statistical characteristics of the velocity field derivatives as,

$$D^s = \frac{1}{2}\sigma_\omega^2 \tau_\omega, \quad D^P = \sigma_{\text{div }\mathbf{u}}^2 \tau_{\text{div }\mathbf{u}}. \tag{4.8}$$

We are interested in two examples of the random velocity field:

(1) incompressible hydrodynamic turbulence;
(2) potential hydrodynamic fields.

A particular case of the potential random field is exemplified by wave turbulence, where the correlation function of the velocity field is given by the following expression

$$B_{ij}(\mathbf{r}, t) = \int d\mathbf{k} \frac{k_i k_j}{k^2} E^P(k) e^{-\lambda k^2 t} \cos\{\mathbf{kr} - \omega(\mathbf{k})t\}, \tag{4.9}$$

where $\omega = \omega(\mathbf{k}) > 0$ defines the dispersion curve for wave motions, and the parameter λ describes wave damping.

The variance of the velocity field in this case takes the form

$$\sigma_{\mathbf{u}}^2 = \langle \mathbf{u}^2(\mathbf{r}, t)\rangle = \int d\mathbf{k} E^P(k), \tag{4.10}$$

and the quantity analogous to that in formula (4.4), is defined as

$$B_{ij}(\mathbf{r}) = \int_0^\infty dt\, B_{ij}(\mathbf{r}, t) = \int d\mathbf{k} \frac{k_i k_j}{k^2} \left[E_1^P(k) \cos \mathbf{kr} + E_2^P(k) \sin \mathbf{kr} \right], \tag{4.11}$$

where

$$E_1^P(k) = E^P(k) \frac{\lambda k^2}{\lambda^2 k^4 + \omega^2(k)}, \quad E_2^P(k) = E^P(k) \frac{\omega(k)}{\lambda^2 k^4 + \omega^2(k)}. \tag{4.12}$$

The detection and description of the phenomenon of spatial structure formation (clustering) in individual realizations of random fields prove to be possible only by analyzing one-time and one-point probability densities of solutions to equations given above if one resorts to the ideas of statistical topography. We consider first the statistical description of lognormal random processes.

Chapter 5
Lognormal Processes, Intermittency, and Dynamic Localization

The one-time probability density $P(y, t; \alpha) = \langle \delta(y(t; \alpha) - y) \rangle$ of lognormal process (2.2) obeys the Fokker–Planck equation

$$\frac{\partial}{\partial t} P(y, t; \alpha) = \left(\alpha \frac{\partial}{\partial y} y + D \frac{\partial}{\partial y} y \frac{\partial}{\partial y} y \right) P(y, t; \alpha),$$
$$P(y, 0; \alpha) = \delta(y - 1),$$

(5.1)

the solution of which, naturally, depends on the parameter α:

$$P(y, t; \alpha) = \frac{1}{2y\sqrt{\pi Dt}} \exp \left\{ -\frac{\ln^2 \left(y e^{\alpha t} \right)}{4Dt} \right\}.$$

(5.2)

Probability distribution (5.2) implies a substantially different behavior for $\alpha > 0$ and $\alpha < 0$. The common feature in both cases is only the appearance of long, moderately sloping 'tails' at large t, indicating an increasing role of large excursions of processes $y(t; \alpha)$ in the formation of one-time statistics. The plots of logarithmically normal probability densities (5.2) for $\alpha > 0$ and $\alpha < 0$ for the parameter $|\alpha|/D = 1$ and dimensionless time $\tau = Dt = 0.1$ and 1 are given in Fig. 5.1. Figure 5.2 shows probability densities of these processes at $\tau = 1$ with log(y) as abscissa.

Accordingly, the integral probability distribution function is given by the expression

$$F(y, t; \alpha) = \int_{-\infty}^{y} dy' P(y', t; \alpha) = \mathrm{P}\left(y(t; \alpha) < y \right) = \Pr \left(\frac{1}{\sqrt{2Dt}} \ln \left(y e^{\alpha t} \right) \right),$$

(5.3)

© Springer International Publishing AG 2017
V.I. Klyatskin, *Fundamentals of Stochastic Nature Sciences*,
Understanding Complex Systems, DOI 10.1007/978-3-319-56922-2_5

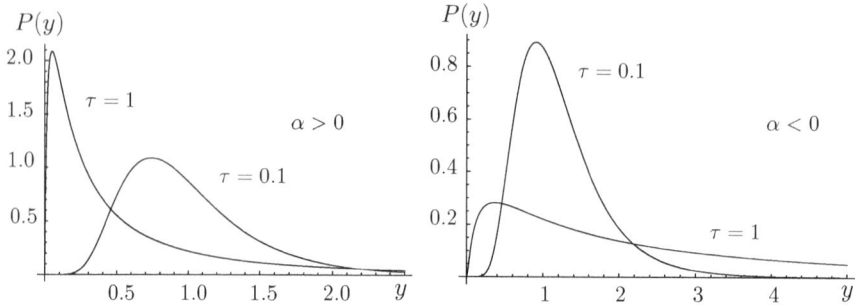

Fig. 5.1 Lognormal probability distributions (5.2) for the parameter $\alpha/D = 1$ and dimensionless time $\tau = 0.1$ and 1

Fig. 5.2 Lognormal probability distributions of random processes $y(t; \alpha)$ for $\alpha > 0$ *solid line* and $\alpha < 0$ *dashed line* at $\tau = 1$ in common logarithmic scale

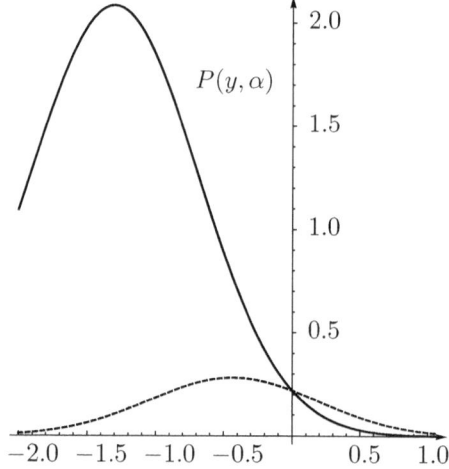

where the function $\Pr(z)$ is the *probability integral* defined as

$$\Pr(z) = \frac{1}{\sqrt{2\pi}} \int_{-\infty}^{z} dx \exp\left\{-\frac{x^2}{2}\right\}. \tag{5.4}$$

Obviously $\Pr(\infty) = 1$ and $\Pr(0) = 1/2$. The asymptotic form of the probability integral for $z \to \pm\infty$ can easily be found from expression (5.4), namely

$$\Pr(z)_{z\to\infty} \approx 1 - \frac{1}{z\sqrt{2\pi}} \exp\left\{-\frac{z^2}{2}\right\}, \quad \Pr(z)_{z\to-\infty} \approx \frac{1}{|z|\sqrt{2\pi}} \exp\left\{-\frac{z^2}{2}\right\}. \tag{5.5}$$

From Eq. (5.1), we can also easily derive the equality

$$\alpha = -\lim_{t \to \infty} \frac{\partial \langle \ln y(t; \alpha) \rangle}{\partial t}. \tag{5.6}$$

It follows then that the parameter α in Eq. (5.1) coincides with the Lyapunov characteristic exponent (5.6) for the lognormal random process $y(t; \alpha)$ (2.2) (see, for example, reviews [55, 56]).

5.1 Typical Realization Curve of a Random Process

The statistical characteristics of the process $z(t)$ at a fixed time instant t are described by the probability density $P(z, t)$ and the integral probability distribution function $F(Z, t) = \int_{-\infty}^{Z} dz' \, P(z', t)$.

The typical realization curve for the random process $z(t)$ is referred to as the deterministic curve $z^*(t)$, which is the *median of the integral probability distribution function* and is defined through the solution of the algebraic equation

$$F\left(z^*(t), t\right) = 1/2. \tag{5.7}$$

This implies, on the one hand, that for any time instant t the probability $\mathsf{P}\{z(t) > z^*(t)\} = \mathsf{P}\{z(t) < z^*(t)\} = 1/2$.

On the other hand, the median has a specific property that for any interval (t_1, t_2) the random process $z(t)$ 'winds round' the curve $z^*(t)$ so that the mean time during which $z(t) > z^*(t)$, coincides with that when the reverse inequality $z(t) < z^*(t)$ holds (Fig. 5.3), i.e. one has

$$\left\langle T_{z(t)>z^*(t)}\right\rangle = \left\langle T_{z(t)<z^*(t)}\right\rangle = \frac{1}{2}\left(t_2 - t_1\right).$$

The curve $z^*(t)$, needless to say, may differ essentially from any individual realization of the process $z(t)$, and does not describe the amplitude of possible excursions.

Fig. 5.3 Regarding the definition of a typical realization curve for a random process

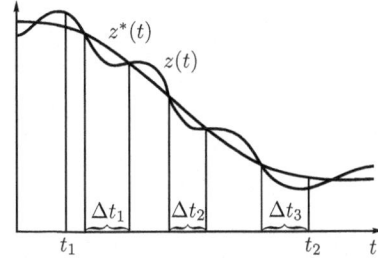

Thus, the typical realization curve $z^*(t)$ of a random process $z(t)$, obtained from the one-time probability density, is defined nevertheless over the entire time interval $t \in (0, \infty)$ and is namely that deterministic curve relative to which the intermittency is enfolding.

The typical realization curve (5.7) for a Gaussian random process $z(t)$ coincides with expectation of the process $z(t)$, i.e., $z^*(t) = \langle z(t) \rangle$, while the typical realization curve for a lognormal process $f(t) = e^{z(t)}$ is defined by the equality $f^*(t) = e^{\langle z(t) \rangle} = e^{\langle \ln f(t) \rangle}$. As a consequence, the typical realization curve of lognormal process (2.2) is described by the formula

$$f^*(t) = e^{\langle \ln f(t) \rangle} = e^{-\alpha t},$$

which coincides with the Lyapunov exponential function.

For $\alpha > 0$ the typical realization curve decays exponentially with time and, in the opposite case, $\alpha < 0$, grows exponentially; namely these functions are plotted in Fig. 2.1 with dashed lines. At $\alpha = 0$ the intermittency takes place with respect to the line $f^*(t) = 1$.

5.2 Dynamic Localization

We note that for *one-dimensional problems the positivity of the Lyapunov character-istic index α corresponds well to the physical phenomenon of dynamic localization* (clustering).

For example, in the problem on diffusion of an ensemble of particles (3.1) in a Gaussian random velocity field $\mathbf{u}(\mathbf{r},t)$, the typical realization curve for the small initial distance between two particles is given by the exponential function of time:

$$l^*(t) = l_0 \exp \left\{ \frac{1}{d(d+2)} \left[D^s d(d-1) - D^P(4-d) \right] t \right\}, \qquad (5.8)$$

where d is the space dimension, and the coefficients D^s and D^P are described by Eq. (4.7) (see, for example, Refs. [28, 29]).

From the last formula, it follows that in the two-dimensional case ($d = 2$) the expression

$$l^*(t) = l_0 \exp \left\{ \frac{1}{4} \left(D^s - D^P \right) t \right\}$$

substantially depends on the sign of the difference ($D^s - D^P$). In particular, for a nondivergent velocity field ($D^P = 0$), we have an exponentially growing typical realization curve, which corresponds to particles running away exponentially fast at small distances between them. In the other limiting case—for a potential velocity field ($D^s = 0$)—the typical realization curve will be an exponentially decaying one; hence, the obvious tendency of particles to 'coalesce'. Thus, the condition for clustering in the two-dimensional case reduces to holding the inequality $D^s < D^P$.

In the three-dimensional case ($d = 3$), Eq. (5.8) gives

$$l^*(t) = l_0 \exp \left\{ \frac{1}{15} \left(6D^s - D^P \right) t \right\},$$

and the typical realization curve will decay with time if the condition: $D^P > 6D^s$, which is more demanding than in the two-dimensional case, is satisfied.

In the one-dimensional case, one finds $l^*(t) = l_0 e^{-D^P t}$, and the typical realization curve always decays with time, for in this case the velocity field is always potential.

For a boundary-value problem (3.12), (3.13) dealing with a plane wave incident on a half-space of a random layered medium, the wave field intensity $I(x) = |u(x)|^2$ is a lognormal random process with the typical realization curve $I^*(x) = 2e^{-D(L-x)}$, where parameter $D = k^2 \sigma_\varepsilon^2 l_0 / 2$ (σ_ε^2 is the variance of process $\varepsilon(x)$, and l_0 is its correlation radius) is the quantity that appears in the correlation function of the Gaussian random process $\varepsilon(x)$ in the delta-correlated approximation with the correlation function $\langle \varepsilon(x)\varepsilon(x') \rangle = 2D\delta(x - x')$.

The coefficient of wave transmission through a sufficiently thick layer of a random medium decays exponentially for the problem considered, and the half-space of a randomly inhomogeneous medium ($L_0 \to -\infty$) completely reflects the wave incident on it. In this case, the wave field intensity $I(x)$ is statistically equivalent to the random process $2y(t; \alpha)$ at $\alpha = D$, and its realization resembles the mirror reflection of Fig. 2.1 ($\alpha > 0$). And yet, certainly, the moments of wave field intensity exponentially grow with distance from the wave source into the medium. We mention that in monograph [47] this effect was established through the analysis of the Lyapunov exponent for the problem at hand, which matches the typical realization curve of the lognormal process.

We also mention that the quantity inverse to the diffusion coefficient D of this problem, defining a natural length scale related to medium random inhomogeneities, is commonly termed the *localization length*, $l_{\mathrm{loc}} = 1/D$.

Later, analyzing statistical characteristics of the wave field, we will see that namely this quantity defines the scale of wave dynamical localization in individual wave field realizations, whereas statistical localization related to wave field statistical characteristics may be absent in a number of cases.

Chapter 6
Stochastic Parametric Resonance

As the first example, consider in more detail stochastic equation of the second order (2.3), which is equivalent to the system of equations of the first order

$$\frac{d}{dt}x(t) = y(t), \quad \frac{d}{dt}y(t) = -2\gamma y(t) - \omega_0^2[1 + z(t)]x(t), \tag{6.1}$$

We will assume that random process $z(t)$ is Gaussian process with parameters

$$\langle z(t)\rangle = 0, \quad \langle z(t)z(t')\rangle = 2\sigma^2\tau_0\delta(t - t').$$

Instead of functions $x(t)$ and $y(t)$ we introduce new ones, namely oscillation amplitude and phase by the formulas

$$x(t) = A(t)\sin(\omega_0 t + \phi(t)), \quad y(t) = \omega_0 A(t)\cos(\omega_0 t + \phi(t)). \tag{6.2}$$

Substituting Eq. (6.2) in the system (6.1), we obtain the system of equations in functions $A(t)$ and $\phi(t)$:

$$\begin{aligned}
\frac{d}{dt}A(t) &= -2\gamma A(t)\cos^2\psi(t) - \frac{\omega_0}{2}z(t)A(t)\sin(2\psi(t)), \\
\frac{d}{dt}\phi(t) &= 2\gamma\sin(2\psi(t)) + \omega_0 z(t)\sin^2\psi(t),
\end{aligned} \tag{6.3}$$

where $\psi(t) = \omega_0 t + \phi(t)$. Using amplitude $A(t)$ in the form $A(t) = e^{u(t)}$, rewrite system (6.3) in the form

$$\begin{aligned}
\frac{d}{dt}u(t) &= -2\gamma\cos^2\psi(t) - \frac{\omega_0}{2}z(t)\sin(2\psi(t)), \\
\frac{d}{dt}\phi(t) &= \gamma\sin(2\psi(t)) + \omega_0 z(t)\sin^2\psi(t).
\end{aligned} \tag{6.4}$$

© Springer International Publishing AG 2017
V.I. Klyatskin, *Fundamentals of Stochastic Nature Sciences*,
Understanding Complex Systems, DOI 10.1007/978-3-319-56922-2_6

Consider now the joint probability density of the solution to the system (6.4) $P(u, \phi, t) = \langle \varphi(u, \phi, t) \rangle$, where the indicator function

$$\varphi(u, \phi, t) = \delta\left(u(t) - u\right) \delta\left(\phi(t) - \phi\right)$$

satisfies the Liouville equation that follows from Eq. (6.4),

$$\frac{\partial}{\partial t} \varphi(u, \phi, t) = \gamma \left\{ 2 \frac{\partial}{\partial u} \cos^2 \psi(t) - \frac{\partial}{\partial \phi} \sin\left(2\psi(t)\right) \right\} \varphi(u, \phi, t)$$

$$+ z(t)\omega_0 \left\{ \frac{1}{2} \frac{\partial}{\partial u} \sin\left(2\psi(t)\right) - \frac{\partial}{\partial \phi} \sin^2 \psi(t) \right\} \varphi(u, \phi, t). \quad (6.5)$$

Averaging then Eq. (6.5) over an ensemble of realizations of random delta-correlated process $z(t)$, we arrive at the Fokker-Plank equation in the probability density

$$\frac{\partial}{\partial t} P(u, \phi, t) = \gamma \left\{ 2 \frac{\partial}{\partial u} \cos^2 \psi(t) - \frac{\partial}{\partial \phi} \sin\left(2\psi(t)\right) \right\} P(u, \phi, t)$$

$$+ D \left\{ \frac{1}{2} \frac{\partial}{\partial u} \sin\left(2\psi(t)\right) - \frac{\partial}{\partial \phi} \sin^2 \psi(t) \right\}^2 P(u, \phi, t),$$

where $D = \sigma^2 \tau_0 \omega_0^2$, which can be rewritten in the form

$$\frac{\partial}{\partial t} P(u, \phi, t) = \gamma \left\{ 2 \frac{\partial}{\partial u} \cos^2 \psi(t) - \frac{\partial}{\partial \phi} \sin\left(2\psi(t)\right) \right\} P(u, \phi, t)$$

$$+ D \left\{ \frac{\partial}{\partial u} \cos\left(2\psi(t)\right) \sin^2 \psi(t) - 2 \frac{\partial}{\partial \phi} \sin^3 \psi(t) \cos \psi(t) \right\} P(u, \phi, t)$$

$$+ D \left\{ \frac{1}{4} \frac{\partial^2}{\partial u^2} \sin^2\left(2\psi(t)\right) - \frac{\partial^2}{\partial u \partial \phi} \sin\left(2\psi(t)\right) \sin^2 \psi(t) + \frac{\partial^2}{\partial \phi^2} \sin^4 \psi(t) \right\}$$

$$\times P(u, \phi, t). \quad (6.6)$$

Assuming absorption parameter sufficiently small, $\gamma \ll \omega_0$, we can average Eq. (6.6) over oscillation period $T = 2\pi/\omega_0$ (in view of the assumption on smallness of variations of statistical characteristics on times (6.6), in view of the assumption on smallness of variations of statistical characteristics on times $\sim T$) to obtain the equation in averaged probability density that describes slow variations of statistical characteristics,

$$\frac{\partial}{\partial t} \overline{P(u, \phi, t)}$$

$$= \gamma \frac{\partial}{\partial u} \overline{P(t; u, \phi)} - \frac{D}{4} \frac{\partial}{\partial u} \overline{P(u, \phi, t)} + \frac{D}{8} \frac{\partial^2}{\partial u^2} \overline{P(t; u, \phi)} + \frac{3D}{8} \frac{\partial^2}{\partial \phi^2} \overline{P(u, \phi, t)}$$

$$(6.7)$$

with the initial condition

$$\overline{P(u, \phi, 0)} = \delta(u - u_0)\delta(\phi - \phi_0).$$

For example, using initial conditions $u_0 = 0$, $\phi_0 = 0$, corresponding to $x(0) = 0$, $y(0) = \omega_0$, we obtain from Eq. (6.7) that statistical characteristics of oscillation amplitude and phase (averaged over the oscillation period) are statistically independent and the corresponding probability densities are the Gaussian ones,

$$\overline{P(u, t)} = \frac{1}{\sqrt{2\pi\sigma_u^2(t)}} \exp\left\{-\frac{(u - \langle u(t)\rangle)^2}{2\sigma_u^2(t)}\right\},$$

$$\overline{P(\phi, t)} = \frac{1}{\sqrt{2\pi\sigma_\phi^2(t)}} \exp\left\{-\frac{(\phi - \phi_0)^2}{2\sigma_\phi^2(t)}\right\},$$

(6.8)

where

$$\langle u(t)\rangle = u_0 - \gamma t + \frac{D}{4}t, \quad \sigma_u^2(t) = \frac{D}{4}t, \quad \langle \phi(t)\rangle = \phi_0, \quad \sigma_\phi^2(t) = \frac{3D}{4}t.$$

The random amplitude is distributed in this case according to the lognormal probability distribution, so that the typical realization curve of the amplitude has the form and exponentially decays

$$A^*(t) = A_0 e^{-(\gamma - D/4)t},$$

under the condition $\gamma > \dfrac{D}{4}$ while statistics of the amplitude $A(t)$ is completely governed by the rare large outliers from this curve. Such a pattern corresponds to the *dynamic localization* of the oscillation amplitude $A(t)$ and follows from its lognormal behavior.

If the above condition is violated, the typical realization curve exponentially increases with time and no dynamical localization occurs.

Chapter 7
Wave Localization in Randomly Layered Media

The problem on plane wave propagation in layered media is formulated in terms of the one-dimensional boundary-value problem. It attracts attention of many researchers because it is much simpler in comparison with the corresponding two- and three-dimensional problems and provides a deep insight into wave propagation in random media. In view of the fact that the one-dimensional problem allows an exact asymptotic solution, we can use it for tracing the effect of different models, medium parameters, and boundary conditions on statistical characteristics of the wavefield (see, for example, the monographs [28, 29]).

The problem in the one-dimensional statement was given in Sect. 3.2.

Three simple statistical problems are of interest:

- Wave incidence on medium layer (of finite and infinite thickness);
- Wave source in the medium layer or infinite medium;
- Effect of boundaries on statistical characteristics of the wavefield.

All these problems can be exhaustively solved in the analytic form. One can easily simulate these problems numerically and compare the simulated and analytic results.

We will assume that $\varepsilon_1(x)$ is the Gaussian delta-correlated random process with the parameters

$$\langle \varepsilon_1(L) \rangle = 0, \quad \langle \varepsilon_1(L)\varepsilon_1(L') \rangle = B_\varepsilon(L - L') = 2\sigma_\varepsilon^2 l_0 \delta(L - L'), \qquad (7.1)$$

where $\sigma_\varepsilon^2 \ll 1$ is the variance and l_0 is the correlation radius of random function $\varepsilon_1(L)$. This approximation means that asymptotic limit process to asymptotic case $l_0 \to 0$ in the exact problem solution with a finite correlation radius l_0 must give the result coinciding with the solution to the statistical problem with parameters (7.1).

In view of smallness of parameter σ_ε^2, all statistical effects can be divided into two types, local and accumulated due to multiple wave reflections in the medium. Our concern will be with the latter.

© Springer International Publishing AG 2017
V.I. Klyatskin, *Fundamentals of Stochastic Nature Sciences*,
Understanding Complex Systems, DOI 10.1007/978-3-319-56922-2_7

The statement of boundary wave problems in terms of the imbedding method clearly shows that two types of wavefield characteristics are of immediate interest. The first type of characteristics deals with quantities, such as values of the wavefield at layer boundaries (reflection and transmission coefficients R_L and T_L), field at the point of source location $G(x_0; x_0)$, and energy flux density at the point of source location $S(x_0; x_0)$. The second type of characteristics deals with statistical characteristics of wavefield intensity in the medium layer, which is the subject matter of the statistical theory of radiative transfer.

7.1   Statistics of Scattered Field at Layer Boundaries

7.1.1   Reflection and Transmission Coefficients

Complex coefficient of wave reflection from a medium layer satisfies the closed Riccati equation (3.21).

Represent reflection coefficient in the form $R_L = \rho_L e^{i\phi_L}$, where ρ_L is the modulus and ϕ_L is the phase. Then, starting from Eq. (3.21), we obtain the system of equations for the squared modulus of the reflection coefficient $W_L = \rho_L^2 = |R_L|^2$ and its phase

$$\frac{d}{dL} W_L = -2k\gamma W_L + k\varepsilon_1(L)\sqrt{W_L}\,(1 - W_L)\sin\phi_L, \quad W_{L_0} = 0,$$

$$\frac{d}{dL}\phi_L = 2k + k\varepsilon_1(L)\left\{1 + \frac{1 + W_L}{2\sqrt{W_L}}\cos\phi_L\right\}, \quad \phi_{L_0} = 0. \tag{7.2}$$

Fast functions producing only little contribution to accumulated effects are omitted in the dissipative terms of system (7.2) (cf. with Eq. (3.23)).

Introduce the indicator function

$$\varphi(L; W) = \delta(W_L - W)$$

that satisfies the Liouville equation

$$\frac{\partial}{\partial L}\varphi(L; W) = 2k\gamma\frac{\partial}{\partial W}\{W\varphi(L; W)\}$$

$$- k\varepsilon_1(L)\frac{\partial}{\partial W}\left\{\sqrt{W}\,(1 - W)\sin\phi_L\varphi(L; W)\right\}. \tag{7.3}$$

Average this equation over an ensemble of realizations of function $\varepsilon_1(L)$. As a result we obtain the unclosed equation for the probability density of reflection coefficient squared modulus $P(L; W) = \langle\varphi(L; W)\rangle$

$$\frac{\partial}{\partial L} P(L, W) = 2k\gamma \frac{\partial}{\partial W} \{W P(L, W)\}$$

$$- k^2 \sigma_\varepsilon^2 l_0 \frac{\partial}{\partial W} (1 - W) \left\langle \left[\sqrt{W} \cos \phi_L + \frac{1}{2} (1 + W) \cos^2 \phi_L \right] \varphi(L, W) \right\rangle$$

$$+ k^2 \sigma_\varepsilon^2 l_0 \frac{\partial}{\partial W} \left\{ \sqrt{W} (1 - W) \frac{\partial}{\partial W} \left[\sqrt{W} (1 - W) \langle \sin^2 \phi_L \varphi(L, W) \rangle \right] \right\}.$$

In view of the fact that the phase of the reflection coefficient $\phi_L = k(L - L_0) + \tilde{\phi}_L$, rapidly varies within distances of about the wavelength, we can additionally average this equation over fast oscillations, which will be valid under the natural restriction $k/D \gg 1$. Thus we arrive at the Fokker–Planck equation

$$\frac{\partial}{\partial L} P(L; W) = 2k\gamma \frac{\partial}{\partial W} W P(L; W) - 2D \frac{\partial}{\partial W} W (1 - W) P(L; W)$$

$$+ D \frac{\partial}{\partial W} W (1 - W)^2 \frac{\partial}{\partial W} P(L; W), \quad P(L_0, W) = \delta (W - 1) \quad (7.4)$$

with the diffusion coefficient

$$D = \frac{k^2 \sigma_\varepsilon^2 l_0}{2}.$$

Representation of quantity W_L in the form

$$W_L = \frac{u_L - 1}{u_L + 1}, \quad u_L = \frac{1 + W_L}{1 - W_L}, \quad u_L \geq 1. \quad (7.5)$$

appears to be more convenient in some cases. Quantity u_L satisfies the stochastic system of equations

$$\frac{d}{dL} u_L = -k\gamma \left(u_L^2 - 1 \right) + k\varepsilon_1(L) \sqrt{u_L^2 - 1} \sin \phi_L, \quad u_{L_0} = 1,$$

$$\frac{d}{dL} \phi_L = 2k + k\varepsilon_1(L) \left\{ 1 + \frac{u_L}{\sqrt{u_L^2 - 1}} \cos \phi_L \right\}, \quad \phi_{L_0} = 0,$$

and we obtain that probability density

$$P(L; u) = \langle \delta(u_L - u) \rangle$$

of random quantity u_L satisfies the Fokker–Planck equation

$$\frac{\partial}{\partial L} P(L; u) = k\gamma \frac{\partial}{\partial u} \left(u^2 - 1\right) P(L; u) + D \frac{\partial}{\partial u} \left(u^2 - 1\right) \frac{\partial}{\partial u} P(L; u). \quad (7.6)$$

Nondissipative Medium

If the medium is non-absorptive (i.e., if $\gamma = 0$), then Eq. (7.6) assumes the form

$$\frac{\partial}{\partial \eta} P(\eta; u) = \frac{\partial}{\partial u} \left(u^2 - 1\right) \frac{\partial}{\partial u} P(\eta; u) \quad (7.7)$$

where we introduced the dimensionless layer thickness $\eta = D(L - L_0)$.

The solution to this equation can be easily obtained using the integral *Meler–Fock transform*

$$P(\eta, u) = \int\limits_0^\infty d\mu\, \mu \tanh(\pi\mu) \exp\left\{-\left(\mu^2 + \frac{1}{4}\right)\eta\right\} P_{-\frac{1}{2}+i\mu}(u), \quad (7.8)$$

where $P_{-1/2+i\mu}(x)$ is the first-order *Legendre function (conal function)*.

In view of the formula

$$\int\limits_1^\infty \frac{dx}{(1+x)^n} P_{-\frac{1}{2}+i\mu}(x) = \frac{\pi}{\cosh(\mu\pi)} K_n(\mu),$$

where

$$K_{n+1}(\mu) = \frac{1}{2n} \left[\mu^2 + \left(n - \frac{1}{2}\right)^2\right] K_n(\mu), \quad K_1(\mu) = 1,$$

representation (7.8) offers a possibility of calculating statistical characteristics of reflection and transmission coefficients $W_L = |R_L|^2$ and $|T_L|^2 = 1 - |R_L|^2 = 2/(1 + u_L)$; in particular, we obtain the expression for the moments of the transmission coefficient squared modulus

$$\langle |T_L|^{2n} \rangle = 2^n \pi \int\limits_0^\infty d\mu \frac{\mu \sinh(\mu\pi)}{\cosh^2(\mu\pi)} K_n(\mu) e^{-(\mu^2+1/4)\eta}. \quad (7.9)$$

Figure 7.1 shows coefficients

$$\langle W_L \rangle = \langle |R_L|^2 \rangle \quad \text{and} \quad \langle |T_L|^2 \rangle = 1 - \langle |R_L|^2 \rangle$$

as functions of layer thickness.

Fig. 7.1 Quantities $\langle |R_L|^2 \rangle$ and $\langle |T_L|^2 \rangle$ versus layer thickness

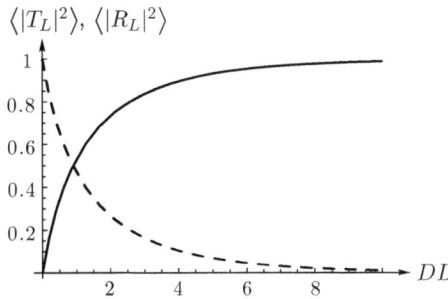

For sufficiently thick layers, namely, if $\eta = D(L - L_0) \gg 1$, Eq. (7.9) yields the asymptotic formula for the moments of the reflection coefficient squared modulus

$$\langle |T_L|^{2n} \rangle \approx \frac{[(2n - 3)!!]^2 \, \pi^2 \sqrt{\pi}}{2^{2n-1}(n - 1)!} \frac{1}{\eta \sqrt{\eta}} e^{-\eta/4}.$$

As may be seen, all moments of the reflection coefficient modulus $|T_L|$ vary with layer thickness according to the universal law (only the numerical factor is changed).

The fact that all moments of quantity $|T_L|$ tend to zero with increasing layer thickness means that $|R_L| \to 1$ with a probability equal to unity, i.e., *the half-space of randomly layered nondissipative medium completely reflects the incident wave.* It is clear that this phenomenon is independent of the statistical model of medium and the condition of applicability of the description based on the additional averaging over fast oscillations related to the reflection coefficient phase.

In the approximation of the delta-correlated random process $\varepsilon_1(L)$, random processes W_L and u_L are obviously the Markovian processes with respect to parameter L. It is obvious that the transition probability density

$$p(u, L|u', L') = \langle \delta(u_L - u|u_{L'} = u') \rangle$$

also satisfies in this case Eq. (7.7), i.e.,

$$\frac{\partial}{\partial L} p(u, L|u', L') = D \frac{\partial}{\partial u} (u^2 - 1) \frac{\partial}{\partial u} p(u, L|u', L')$$

with the initial condition $p(u, L'|u', L') = \delta(u - u')$. The corresponding solution has the form

$$p(u, L|u', L') = \int_0^\infty d\mu \, \mu \tanh(\pi\mu) e^{-D(\mu^2 + 1/4)(L - L')} P_{-\frac{1}{2} + i\mu}(u) P_{-\frac{1}{2} + i\mu}(u'). \quad (7.10)$$

At $L' = L_0$ and $u' = 1$, expression (7.10) grades into the one-point probability density (7.8).

Dissipative Medium

In the case of an absorptive medium, Eqs. (7.4) and (7.6) cannot be solved analytically for the layer of finite thickness. Nevertheless, in the limit of half-space $(L_0 \to -\infty)$, quantities W_L and u_L have the steady-state probability density independent of L and satisfying the equations

$$2(\beta - 1 + W) P(W) + (1 - W)^2 \frac{d}{dW} P(W) = 0, \quad 0 < W < 1,$$

$$\beta P(u) + \frac{d}{du} P(u) = 0, \quad u > 1, \tag{7.11}$$

where $\beta = k\gamma/D$ is the dimensionless absorption coefficient.

Solutions to Eq. (7.11) have the form

$$P(W) = \frac{2\beta}{(1-W)^2} \exp\left\{-\frac{2\beta W}{1-W}\right\}, \quad P(u) = \beta e^{-\beta(u-1)} \tag{7.12}$$

and Fig. 7.2 shows function $P(W)$ for different values of parameter β.

The physical meaning of probability density (7.12) is obvious. It describes the statistics of the reflection coefficient of the random layer sufficiently thick for the incident wave could not reach its end because of dynamic absorption in the medium.

Using distributions (7.12), we can calculate all moments of quantity $W_L = |R_L|^2$. For example, the average square of reflection coefficient modulus is given by the formula

$$\langle W \rangle = \int_0^1 dW\, W P(W) = \int_1^\infty du \frac{u-1}{u+1} P(u) = 1 + 2\beta e^{2\beta}\, \mathrm{Ei}(-2\beta),$$

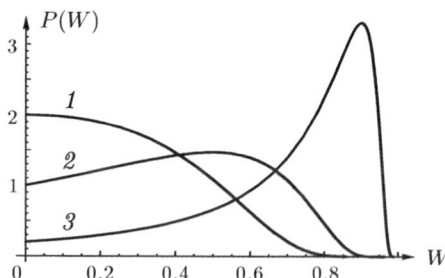

Fig. 7.2 Probability density of squared reflection coefficient modulus $P(W)$. *Curves 1–3* correspond to $\beta = 1, 0.5$, and 0.1, respectively

where $\mathrm{Ei}(-x) = -\int\limits_{x}^{\infty} \dfrac{dt}{t} e^{-t}$ $(x > 0)$ is the *integral exponent*. Using asymptotic expansions of function $\mathrm{Ei}(-x)$

$$\mathrm{Ei}(-x) = \begin{cases} \ln x & (x \ll 1), \\ -e^{-x}\dfrac{1}{x}\left(1 - \dfrac{1}{x}\right) & (x \gg 1), \end{cases}$$

we obtain the asymptotic expansions of quantity $\langle W \rangle = \langle |R_L|^2 \rangle$

$$\langle W \rangle \approx \begin{cases} 1 - 2\beta \ln(1/\beta), & \beta \ll 1, \\ 1/(2\beta), & \beta \gg 1. \end{cases} \tag{7.13}$$

To determine higher moments of quantity $W_L = |R_L|^2$, we multiply the first equation in (7.11) by W^n and integrate the result over W from 0 to 1. As a result, we obtain recurrence equation

$$n\langle W^{n+1} \rangle - 2(\beta + n)\langle W^n \rangle + n\langle W^{n-1} \rangle = 0 \quad (n = 1, 2, \cdots). \tag{7.14}$$

Using this equation, we can recursively calculate all higher moments. For example, we have for $n = 1$

$$\langle W^2 \rangle = 2(\beta + 1)\langle W \rangle - 1.$$

The steady-state probability distribution can be obtained not only by limiting process $L_0 \to -\infty$, but also $L \to \infty$. Equation (7.4) was solved numerically for two values $\beta = 1.0$ and $\beta = 0.08$ for different initial conditions. Figure 7.3 shows moments $\langle W_L \rangle$, $\langle W_L^2 \rangle$ calculated from the obtained solutions versus dimensionless layer thickness $\eta = D(L - L_0)$.

Fig. 7.3 Statistical characteristics of quantity $W_L = |R_L|^2$. *Curves 1 and 2* show the second and first moments at $\beta = 1$, and *curves 3 and 4* show the second and first moments at $\beta = 0.08$

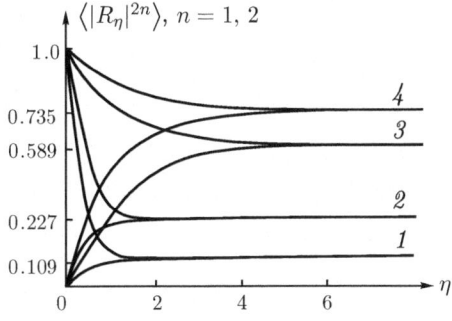

The curves show that the probability distribution approaches its steady-state behavior relatively rapidly ($\eta \sim 1.5$) for $\beta \geq 1$ and much slower ($\eta \geq 5$) for strongly stochastic problem with $\beta = 0.08$.

Note that, in the problem under consideration, energy flux density and wavefield intensity at layer boundary $x = L$ can be expressed in terms of the reflection coefficient. Consequently, we have for $\beta \ll 1$

$$\langle S(L, L) \rangle = 1 - \langle W_L \rangle = 2\beta \ln(1/\beta), \quad \langle I(L, L) \rangle = 1 + \langle W_L \rangle = 2. \quad (7.15)$$

Taking into account that $|T_L| = 0$ in the case of random half-space and using Eq. (3.15), we obtain that the wavefield energy contained in this half-space

$$E = D \int\limits_{-\infty}^{L} dx I(x; L),$$

has the probability distribution

$$P(E) = \beta P(W)|_{W=(1-\beta E)} = \frac{2}{E^2} \exp\left\{-\frac{2}{E}(1 - \beta E)\right\} \theta(1 - \beta E), \quad (7.16)$$

so that we have, in particular, for $\beta \ll 1$

$$\langle E \rangle = 2 \ln(1/\beta). \quad (7.17)$$

Note that probability distribution (7.16) allows the limit process to $\beta = 0$; as a result, we obtain the limiting probability density

$$P(E) = \frac{2}{E^2} \exp\left\{-\frac{2}{E}\right\} \quad (7.18)$$

that decays according to the power law for large energies E. The corresponding integral distribution function has the form

$$F(E) = \exp\left\{-\frac{2}{E}\right\}.$$

A consequence of Eq. (7.18) is the fact that all moments of the total wave energy appear infinite. Nevertheless, the total energy in separate wavefield realizations can be limited to arbitrary value with a finite probability.

7.1.2 Source Inside the Layer of a Medium

If the source of plane waves is located inside the medium layer, the wavefield and energy flux density at the point of source location are given by Eqs. (3.34) and (3.36). Quantities $R_1(x_0)$ and $R_2(x_0)$ are statistically independent within the framework of the model of the delta-correlated fluctuations of $\varepsilon_1(x)$, because they satisfy dynamic equations for nonoverlapping space portions. In the case of the infinite space ($L_0 \rightarrow -\infty, L \rightarrow \infty$), probability densities of quantities $R_1(x_0)$ and $R_2(x_0)$ are given by Eq. (7.12); as a result, average intensity of the wavefield and average energy flux density at the point of source location are given by the expressions

$$\langle I(x_0; x_0) \rangle = 1 + \frac{1}{\beta}, \quad \langle S(x_0; x_0) \rangle = 1. \tag{7.19}$$

The infinite increase of average intensity at the point of source location for $\beta \rightarrow 0$ is evidence of wave energy accumulation in a randomly layered medium; at the same time, average energy flux density at the point of source location is independent of medium parameter fluctuations and coincides with energy flux density in free space.

For the source located at perfectly reflecting boundary $x_0 = L$, we obtain from Eqs. (3.34) and (3.36)

$$\langle I_{\text{ref}}(L; L) \rangle = 4 \left(1 + \frac{2}{\beta} \right), \quad \langle S_{\text{ref}}(L; L) \rangle = 4, \tag{7.20}$$

i.e., average energy flux density of the source located at the reflecting boundary is also independent of medium parameter fluctuations and coincides with energy flux density in free space.

Note the singularity of the above formulas (7.19), (7.20) for $\beta \rightarrow 0$, which shows that absorption (even arbitrarily small) serves the regularizing factor in the problem on the point source.

Using Eq. (3.37), we can obtain the probability distribution of wavefield energy in the half-space

$$E = D \int\limits_{-\infty}^{x_0} dx I(x; x_0).$$

In particular, for the source located at reflecting boundary, we obtain the expression

$$P_{\text{ref}}(E) = \sqrt{\frac{2}{\pi}} \frac{1}{E\sqrt{E}} \exp\left\{ -\frac{2}{E} \left(1 - \frac{\beta E}{4} \right)^2 \right\},$$

that allows limiting process $\beta \rightarrow 0$, which is similar to the case of the wave incidence on the half-space of random medium.

7.1.3 Statistical Localization

In view of Eq. (3.37), the obtained results related to wave field at fixed spatial points (at layer boundaries and at the point of source location) offer a possibility of making certain general conclusions about the behavior of the wavefield average intensity inside the random medium.

For example, Eq. (3.37) yields the expression for average energy contained in the half-space $(-\infty, x_0)$

$$\langle E \rangle = D \int_{-\infty}^{x_0} dx \, \langle I(x; x_0) \rangle = \frac{1}{\beta} \langle S(x_0; x_0) \rangle . \qquad (7.21)$$

In the case of the plane wave $(x_0 = L)$ incident on the half-space $x \le L$, Eqs. (7.15) and (7.21) result for $\beta \ll 1$ in the expressions

$$\langle E \rangle = 2 \ln(1/\beta), \quad \langle I(L; L) \rangle = 2. \qquad (7.22)$$

Consequently, the space segment of length

$$Dl_\beta \cong \ln(1/\beta),$$

concentrates the most portion of average energy, which means that there occurs the *wavefield statistical localization* caused by wave absorption. Note that, in the absence of medium parameter fluctuations, energy localization occurs on scales about absorption length $Dl_{abs} \cong 1/\beta$. However, we have $l_{abs} \gg l_\beta$ for $\beta \ll 1$. If $\beta \to 0$, then $l_\beta \to \infty$, and statistical localization of the wavefield disappears in the limiting case of non-absorptive medium.

In the case of the source in unbounded space, we have

$$\langle E \rangle = \frac{1}{\beta}, \quad \langle I(x_0; x_0) \rangle = 1 + \frac{1}{\beta},$$

and average energy localization is characterized, as distinct from the foregoing case, by spatial scale $D|x - x_0| \cong 1$ for $\beta \to 0$.

In a similar way, we have for the source located at reflecting boundary

$$\langle E \rangle = \frac{4}{\beta}, \quad \langle I_{ref}(L; L) \rangle = 4 \left(1 + \frac{2}{\beta} \right),$$

from which follows that average energy localization is characterized by a half spatial scale $D(L - x)| \cong 1/2$ for $\beta \to 0$.

In the considered problems, wavefield average energy essentially depends on parameter β and tends to infinity for $\beta \to 0$. However, this is the case only for average quantities. In our further analysis of the wavefield in random medium, we will show that the field in a separate realization is localized due to the *dynamic localization* even in non-absorptive media.

7.2 Statistical Theory of Radiative Transfer

Now, we dwell on the statistical description of a wavefield in random medium (statistical theory of radiative transfer). We consider two problems of which the first concerns waves incident on the medium layer and the second concerns waves generated by a source located in the medium.

7.2.1 Normal Wave Incidence on the Layer of Random Media

In the general case of absorptive medium, the wavefield is described by the boundary-value problem (3.12), (3.13). We introduce complex opposite waves

$$u(x) = u_1(x) + u_2(x), \quad \frac{d}{dx}u(x) = -ik[u_1(x) - u_2(x)],$$

related to the wavefield through the relationships

$$u_1(x) = \frac{1}{2}\left[1 + \frac{i}{k}\frac{d}{dx}\right]u(x), \quad u_1(L) = 1,$$

$$u_2(x) = \frac{1}{2}\left[1 - \frac{i}{k}\frac{d}{dx}\right]u(x), \quad u_2(L_0) = 0,$$

so that the boundary-value problem (3.12), (3.13) can be rewritten in the form

$$\left(\frac{d}{dx} + ik\right)u_1(x) = -\frac{ik}{2}\varepsilon(x)[u_1(x) + u_2(x)], \quad u_1(L) = 1,$$

$$\left(\frac{d}{dx} - ik\right)u_2(x) = -\frac{ik}{2}\varepsilon(x)[u_1(x) + u_2(x)], \quad u_2(L_0) = 0.$$

The wavefield as a function of parameter L satisfies imbedding equation (3.22). It is obvious that the opposite waves will also satisfy Eq. (3.22), but with different initial conditions:

$$\frac{\partial}{\partial L} u_1(x; L) = ik \left(1 + \frac{1}{2} \varepsilon(L) (1 + R_L) \right) u_1(x; L), \quad u_1(x; x) = 1,$$

$$\frac{\partial}{\partial L} u_2(x; L) = ik \left(1 + \frac{1}{2} \varepsilon(L) (1 + R_L) \right) u_2(x; L), \quad u_2(x; x) = R_x,$$

where reflection coefficient R_L satisfies Eq. (3.21).

Introduce now opposite wave intensities

$$W_1(x; L) = |u_1(x; L)|^2 \quad \text{and} \quad W_2(x; L) = |u_2(x; L)|^2.$$

They satisfy the equations

$$\frac{\partial}{\partial L} W_1(x; L) = -k\gamma W_1(x; L) + \frac{ik}{2} \varepsilon(L) \left(R_L - R_L^* \right) W_1(x; L),$$

$$\frac{\partial}{\partial L} W_2(x; L) = -k\gamma W_2(x; L) + \frac{ik}{2} \varepsilon(L) \left(R_L - R_L^* \right) W_2(x; L), \tag{7.23}$$

$$W_1(x; x) = 1, \quad W_2(x; x) = |R_x|^2.$$

Quantity $W_L = |R_L|^2$ appeared in the initial condition of Eq. (7.23) satisfies Eq. (3.23) or the equation

$$\frac{d}{dL} W_L = -2k\gamma W_L - \frac{ik}{2} \varepsilon_1(L) \left(R_L - R_L^* \right) (1 - W_L), \quad W_{L_0} = 0. \tag{7.24}$$

In Eqs. (7.23) and (7.24), we omitted dissipative terms producing no contribution in accumulated effects.

As earlier, we will assume that $\varepsilon_1(x)$ is the Gaussian delta-correlated process with correlation function (7.1). In view of the fact that Eqs. (7.23), (7.24) are the first-order equations with initial conditions, we can use the standard procedure of deriving the Fokker–Planck equation for the joint probability density of quantities $W_1(x; L)$, $W_2(x; L)$, and W_L

$$P(x; L; W_1, W_2, W) = \langle \delta(W_1(x; L) - W_1)\delta(W_2(x; L) - W_2)\delta(W_L - W) \rangle.$$

As a result, we obtain the Fokker–Planck equation

$$\frac{\partial}{\partial L} P(x; L; W_1, W_2, W)$$

$$= k\gamma \left(\frac{\partial}{\partial W_1} W_1 + \frac{\partial}{\partial W_2} W_2 + 2 \frac{\partial}{\partial W} W \right) P(x; L; W_1, W_2, W)$$

$$+ D \left[\frac{\partial}{\partial W_1} W_1 + \frac{\partial}{\partial W_2} W_2 - \frac{\partial}{\partial W} (1 - W) \right] P(x; L; W_1, W_2, W)$$

$$+ D \left[\frac{\partial}{\partial W_1} W_1 + \frac{\partial}{\partial W_2} W_2 - \frac{\partial}{\partial W} (1 - W) \right]^2 W P(x; L; W_1, W_2) \quad (7.25)$$

with the initial condition

$$P(x; x; W_1, W_2, W) = \delta(W_1 - 1)\delta(W_2 - W)P(x; W),$$

where function $P(L; W)$ is the probability density of reflection coefficient squared modulus W_L, which satisfies Eq. (7.4). As earlier, the diffusion coefficient in Eq. (7.25) is $D = k^2 \sigma_\varepsilon^2 l_0 / 2$. Deriving this equation, we used an additional averaging over fast oscillations ($u(x) \sim e^{\pm ikx}$) that appear in the solution of the problem at $\varepsilon = 0$.

In view of the fact that Eq. (7.23) are linear in $W_n(x; L)$, we can introduce the generating function of moments of opposite wave intensities

$$Q(x; L; \mu, \lambda, W) = \int_0^1 dW_1 \int_0^1 dW_2 \, W_1^{\mu - \lambda} W_2^\lambda P(x; L; W_1, W_2, W), \quad (7.26)$$

which satisfies the simpler equation

$$\frac{\partial}{\partial L} Q(x; L; \mu, \lambda, W) = -k\gamma \left(\mu - 2\frac{\partial}{\partial W} W \right) Q(x; L; \mu, \lambda, W)$$

$$- D \left[\mu + \frac{\partial}{\partial W} (1 - W) \right] Q(x; L; \mu, \lambda, W)$$

$$+ D \left[\mu - \frac{\partial}{\partial W} (1 - W) \right]^2 W Q(x; L; \mu, \lambda, W) \quad (7.27)$$

with the initial condition

$$Q(x; x; \mu, \lambda, W) = W^\lambda P(x; W).$$

In terms of function $Q(x; L; \mu, \lambda, W)$, we can determine the moment functions of opposite wave intensities by the formula

$$\left\langle W_1^{\mu - \lambda}(x; L) W_2^\lambda(x; L) \right\rangle = \int_0^1 dW \, Q(x; L; \mu, \lambda, W). \quad (7.28)$$

Equation (7.27) describes statistics of the wavefield in medium layer $L_0 \leq x \leq L$. In particular, at $x = L_0$, it describes the transmission coefficient of the wave.

In the limiting case of the half-space ($L_0 \to -\infty$), Eq. (7.27) grades into the equation

$$\frac{\partial}{\partial \xi} Q(\xi; \mu, \lambda, W) = -\beta \left(\mu - 2\frac{\partial}{\partial W} W \right) Q(\xi; \mu, \lambda, W)$$

$$- \left[\mu + \frac{\partial}{\partial W} (1 - W) \right] Q(\xi; \mu, \lambda, W)$$

$$+ \left[\mu - \frac{\partial}{\partial W} (1 - W) \right]^2 W Q(\xi; \mu, \lambda, W) \quad (7.29)$$

with the initial condition

$$Q(0; \mu, \lambda, W) = W^\lambda P(W),$$

where $\xi = D(L - x) > 0$ is the dimensionless distance, and steady-state (independent of L) probability density of the reflection coefficient modulus $P(W)$ is given by Eq. (7.12). In this case, Eq. (7.28) assumes the form

$$\left\langle W_1^{\mu - \lambda}(\xi) W_2^\lambda(\xi) \right\rangle = \int_0^1 dW Q(\xi; \mu, \lambda, W). \quad (7.30)$$

Further discussion will be more convenient if we consider separately the cases of absorptive (dissipative) and non-absorptive (nondissipative) random media.

7.2.2 Nondissipative Medium (Stochastic Wave Parametric Resonance and Dynamic Wave Localization)

For non-absorptive medium, imbedding Eqs. (3.21) and (3.22) are simplified. In this case, Eq. (3.23) for wavefield intensity can be integrated analytically, and relationship (3.27) expresses the intensity immediately in terms of the reflection coefficient. Using reflection coefficient in representation (7.5), we can rewrite this relationship in the form

$$\frac{1}{2} I(x, L) = \frac{u_x + \sqrt{u_x^2 - 1} \cos \phi_x}{1 + u_L}, \quad (7.31)$$

where phase ϕ_x of the reflection coefficient has the form $\phi_x = 2kx + \widetilde{\phi}_x$ and u_x and $\widetilde{\phi}_x$ are slow functions on distances of about the wavelength. For this reason, it is expedient to consider only slow variations of combinations of function $I(x, L)$ with respect to x, which corresponds to preliminary averaging the functions that rapidly vary within scales of about the wavelength. We will use overbar to denote such averaging. For example, averaging of Eq. (7.31) gives

$$\frac{1}{2}\overline{I(x, L)} = \frac{u_x}{1 + u_L}. \quad (7.32)$$

We have similarly

$$\frac{1}{4}\overline{I^2(x, L)} = \frac{3u_x^2 - 1}{2(1 + u_L)^2},$$ (7.33)

and so on.

As was mentioned earlier, function u_x appeared in equations like Eqs. (7.32) and (7.33) is the Markovian random process with transition probability density (7.10) and one-point probability density (7.8). Consequently, determination of statistical characteristics of wave intensity reduces simply to calculating a quadrature. For example, for quantity $\overline{I^n(x, L)}$, we obtain the expression

$$\frac{1}{2^n}\overline{I^n(x, L)} = \frac{g_n(u_x)}{(1 + u_L)^n},$$

where $g_n(u_x)$ is the polynomial of power n in u_x, so that

$$\frac{1}{2^n}\left\langle\overline{I^n(x, L)}\right\rangle = \int\limits_1^\infty \frac{du_L}{(1 + u_L)^n} \int\limits_1^\infty du_x g_n(u_x) p(u_L, L|u_x, x) P(x, u_x).$$ (7.34)

Substituting Eq. (7.10) for $p(u_L, L|u_x, x)$ in Eq. (7.34) and using formula

$$\int\limits_1^\infty \frac{dx}{(1 + x)^n} P_{-\frac{1}{2}+i\mu}(x) = \frac{\pi}{\cosh(\mu\pi)} K_n(\mu),$$ (7.35)

where

$$K_{n+1}(\mu) = \frac{1}{2n}\left[\mu^2 + \left(n - \frac{1}{2}\right)^2\right] K_n(\mu), \quad K_1(\mu) = 1,$$

we can perform integration over u_L to obtain the two-fold (in appearance) integral

$$\frac{1}{2^n}\left\langle\overline{I^n(x, L)}\right\rangle = \pi \int\limits_0^\infty d\mu\, \mu \frac{\sinh(\mu\pi)}{\cosh^2(\mu\pi)} K_n(\mu) e^{-(\mu^2+\frac{1}{4})(L-x)}$$

$$\times \int\limits_1^\infty du\, g_n(u) P_{-\frac{1}{2}+i\mu}(u) P(x, u).$$ (7.36)

Here, we introduced dimensionless distances $DL \to L$ and $Dx \to x$. In addition, we assumed that $L_0 = 0$.

In view of the expression

$$I(0; L) = |T_L|^2 = \frac{2}{1 + u_L},$$

the integral

$$\int\limits_{1}^{\infty} \frac{du_L}{(1+u_L)^n} \int\limits_{1}^{\infty} du_x g_k(u_x) p(u_L, L|u_x, x) P(x, u_x)$$

describes correlations of the wave transmission coefficient with the wave intensity in the layer.

Our further task consists in calculating the inner integral in Eq. (7.36), which reduces to the solution of a simple system of differential equations.

Indeed, consider the expressions

$$f_k(x) = \int\limits_{1}^{\infty} du u^k P_{-1/2+i\mu}(u) P(x, u) \quad (k = 0, 1, \cdots), \qquad (7.37)$$

which are the Meler–Fock transforms of functions $u^k P(x; u)$. Differentiating Eq. (7.37) with respect to x, using the Fokker–Planck equation (7.7) for function $P(x; u)$ and differential equation for the *Legendre function* $P_{-1/2+i\mu}(x)$

$$\frac{d}{dx}\left(x^2 - 1\right)\frac{d}{dx}P_{-1/2+i\mu}(x) = -\left(\mu^2 + \frac{1}{4}\right)P_{-1/2+i\mu}(x),$$

and integrating the result by parts, we arrive at the equation

$$\frac{d}{dx}f_k(x) = -\left(\mu^2 + \frac{1}{4} - k^2 - k\right)f_k(x) + 2k\psi_k(x) - k(k-1)f_{k-2}(x), \quad (7.38)$$

where

$$\psi_k(x) = \int\limits_{1}^{\infty} du u^{k-1} P(x, u)\left(u^2 - 1\right)\frac{d}{du}P_{-1/2+i\mu}(u). \qquad (7.39)$$

Differentiating now function $\psi_k(x)$ with respect to x, we similarly obtain that this function satisfies the equation

$$\frac{d}{dx}\psi_k(x) = -\left(\mu^2 + \frac{1}{4} - k^2 + k\right)\psi_k(x) - 2k\left(\mu^2 + \frac{1}{4}\right)f_k(x)$$

$$- (k-1)(k-2)\psi_{k-2}(x) + 2(k-1)\left(\mu^2 + \frac{1}{4}\right)f_{k-2}(x). \quad (7.40)$$

The initial values for Eqs. (7.38) and (7.40) are, obviously, the conditions

$$f_k(0) = 1, \quad \psi_k(0) = 0.$$

Thus, functions $f_k(x)$ and $\psi_k(x)$ are mutually related and satisfy the closed recursive system of inhomogeneous second-order differential equations with constant coefficients, and this system can be easily solved.

Represent the solution to system (7.38), (7.40) in the form

$$f_k(x) = \widetilde{f}_k(x)e^{-(\mu^2+\frac{1}{4}-k^2)x}, \quad \psi_k(x) = \widetilde{\psi}_k(x)e^{-(\mu^2+\frac{1}{4}-k^2)x}. \tag{7.41}$$

Then, for functions $\widetilde{f}_k(x)$ and $\widetilde{\psi}_k(x)$, we obtain the system of equations

$$\left(\frac{d}{dx} - k\right)\widetilde{f}_k(x) = 2k\widetilde{\psi}_k(x) - k(k-1)\widetilde{f}_{k-2}(x)e^{-4(k-1)x},$$

$$\left(\frac{d}{dx} + k\right)\widetilde{\psi}_k(x) = -2k\left(\mu^2 + \frac{1}{4}\right)\widetilde{f}_k(x)$$

$$+ (k-1)\left[2\left(\mu^2 + \frac{1}{4}\right)\widetilde{f}_{k-2}(x) - (k-2)\widetilde{\psi}_{k-2}(x)\right]e^{-4(k-1)x} \tag{7.42}$$

with the initial values
$$\widetilde{f}_k(0) = 1, \quad \widetilde{\psi}_k(0) = 0.$$

We note that the corresponding solution to the homogeneous system has the form

$$\widetilde{f}_k(x) = A(\mu)\sin(2k\mu x) + B(\mu)\cos(2k\mu x).$$

Consider the simplest cases.
1. In the case $k = 0$, we have

$$\frac{d}{dx}\widetilde{f}_0(x) = 0, \quad \widetilde{f}_0(L_0) = 1,$$

so that

$$f_0(x) = \exp\left\{-\left(\mu^2 + \frac{1}{4}\right)x\right\}.$$

Then, the integral

$$\langle|T_L|^{2n}\rangle = \int\limits_1^\infty \frac{du_L}{(1+u_L)^n} \int\limits_1^\infty dup(u_L, L|u, x)P(x, u)$$

$$= 2^n\pi \int\limits_0^\infty d\mu \frac{\mu\sinh(\mu\pi)}{\cosh^2(\mu\pi)}K_n(\mu)e^{-(\mu^2+\frac{1}{4})(L-L_0)}$$

describes the moments of the modulus of coefficient of wave transmission through the layer of random medium.

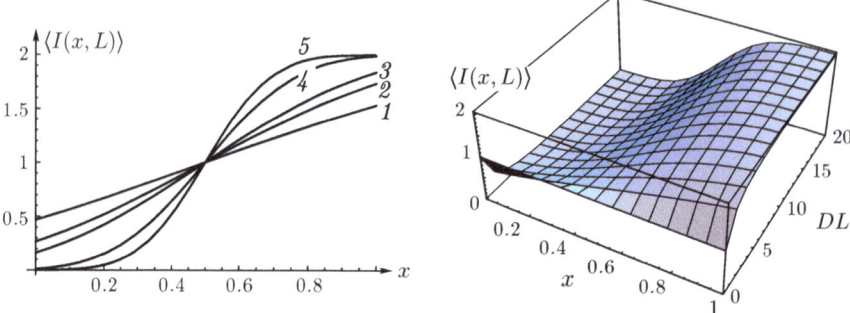

Fig. 7.4 Wavefield average intensity in the problem on a wave incident on medium layer. *Curves* *1–5* correspond to parameter $DL = 1, 2, 3, 10,$ and 20, respectively

2. In the case $k = 1$, we have the system of equations

$$\left(\frac{d}{dx} - 2\right) \tilde{f}_1(x) = 2\tilde{\psi}_1(x), \quad \left(\frac{d}{dx} + 1\right) \tilde{\psi}_1(x) = -2\left(\mu^2 + \frac{1}{4}\right) \tilde{f}_1(x)$$

so that

$$f_1(x) = \exp\left\{-\left(\mu^2 - \frac{3}{4}\right)x\right\} \left(\cos(2\mu x) + \frac{1}{2\mu}\sin(2\mu x)\right).$$

In this case, integral (7.34) at $n = 1$ describes the distribution of the wavefield average intensity in the layer of random medium

$$\langle\overline{I(x, L)}\rangle = 2\pi \int\limits_0^\infty d\mu \frac{\mu \sinh(\mu\pi)}{\cosh^2(\mu\pi)} e^{x - (\mu^2 + \frac{1}{4})L} \left(\cos(2\mu x) + \frac{1}{2\mu}\sin(2\mu x)\right).$$

Figure 7.4 shows this intensity distribution for different layer thicknesses.
3. In the case $k = 2$, we have the system of equations

$$\left(\frac{d}{dx} - 2\right) \tilde{f}_2(x) = 4\tilde{\psi}_2(x) - 2e^{-4x},$$

$$\left(\frac{d}{dx} + 2\right) \tilde{\psi}_2(x) = -2\left(\mu^2 + \frac{1}{4}\right) \left[2\tilde{f}_2(x) - e^{-4x}\right],$$

so that

$$\tilde{f}_2(x) = \frac{\mu^2 + 5/4}{2\left(1 + \mu^2\right)} \cos(4\mu\pi) + \frac{\mu^2 + 3/4}{2\mu\left(1 + \mu^2\right)} \sin(4\mu\pi) + \frac{\mu^2 + 3/4}{2\left(1 + \mu^2\right)} e^{-4x}.$$

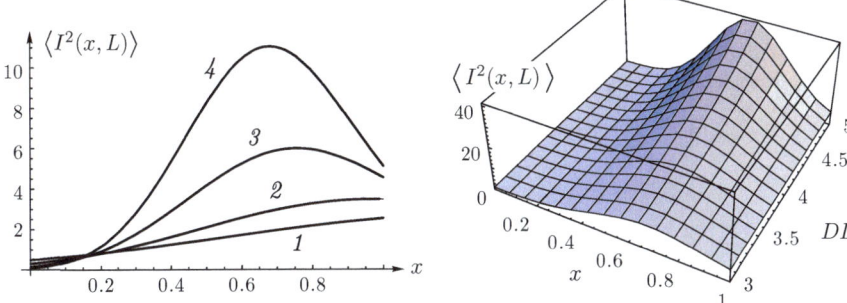

Fig. 7.5 Second moment of wavefield intensity in the problem on a wave incident on medium layer. *Curves 1–4* correspond to parameter $DL = 0.5, 1, 2$, and 3, respectively

In this case, integral (7.36) at $n = 2$ describes the distribution of the second moment of the intensity along the layer

$$\langle \overline{I^2(x, L)} \rangle = \pi \int\limits_0^\infty d\mu \frac{\mu \sinh(\mu\pi)}{\cosh^2(\mu\pi)} e^{-(\mu^2+\frac{1}{4})L} \left(\mu^2 + \frac{1}{4}\right) \left(3 e^{4x} \widetilde{f}_2(x) - 1\right).$$

Figure 7.5 shows this distribution for different layer thicknesses.

Thus, solving successively the recurrent system of equations (7.42), we can express the corresponding moment of intensity in terms of a sole quadrature.

Consider the structure of the obtained expressions. As we have seen earlier, moments of the wavefield intensity in the layer of medium are expressed in terms of the integrals

$$\langle \overline{I^n(x, L)} \rangle \sim \int\limits_{-\infty}^\infty d\mu \frac{\sinh(\mu\pi)}{\cosh^2(\mu\pi)} \Phi(\mu) e^{n^2 x + 2in\mu x - (\mu^2 + \frac{1}{4})L}$$

$$= e^{-\frac{1}{4}L + n^2 L\xi(1-\xi)} \int\limits_{-\infty}^\infty d\mu \frac{\sinh(\mu\pi)}{\cosh^2(\mu\pi)} \Phi(\mu) e^{-(\mu - in\xi)^2 L}, \qquad (7.43)$$

where $\xi = x/L$ and $\Phi(\mu)$ is the algebraic function of parameter μ. If we consider asymptotic limit $L \to \infty$ under the condition that ξ remains finite, then we obtain from Eq. (7.43) that two spatial scales

$$\xi_1 = \frac{1}{2}\left(1 - \sqrt{1 - \frac{1}{n^2}}\right) \text{ and } \xi_2 = 1 - \frac{1}{2n}$$

Fig. 7.6 Schematic of the
behavior of moments of
wavefield intensity in the
problem on a wave incident
on medium layer (Stochastic
parametric resonance)

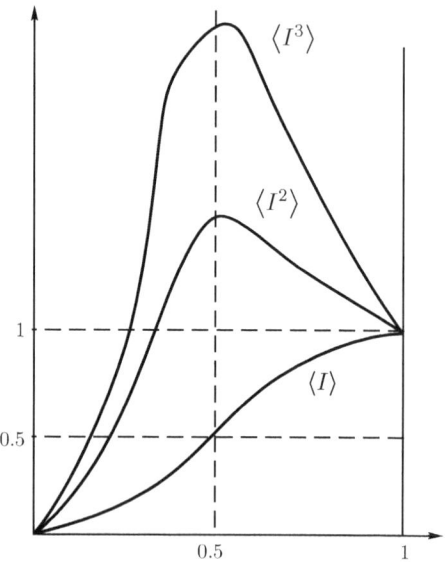

exist such that quantity $\left\langle \overline{I^n(x, L)} \right\rangle$ is exponentially small for $0 \le \xi \le \xi_1$. For $\xi_1 \le \xi \le \xi_2$, quantity $\left\langle \overline{I^n(x, L)} \right\rangle$ is exponentially great and achieves its maximum in the vicinity of point $\xi \approx 1/2$, where $\left\langle \overline{I^n(x, L)} \right\rangle_{\max} \sim \exp\left\{ (n^2 - 1) \, l/4 \right\}$. For $1 \ge \xi \ge \xi_2$, quantity $\left\langle \overline{I^n(x, L)} \right\rangle$ exponentially tends to unity. The above behavior is pertinent to the case $n \ge 2$. The case $n = 1$ forms the exception; in this case, points ξ_1 and ξ_2 merge, and average intensity distribution appears to be monotonic.

The first scale follows from the relationship $n^2 \xi \, (1 - \xi) \sim 1/4$, and the second scale follows from the fact that, in view of limiting condition $\left\langle \overline{I^n(x, L)} \right\rangle \to 2^n$ for $L \to \infty$, integral (7.43) is contributed mainly by the pole $\mu_n = i \, (n - 1/2)$, so that integration contour must run above μ_n, i.e., $\mu_n < i n \xi$. With increasing n, variable $\xi_1 \to 0$, and variable $\xi_2 \to 1$ (see Fig. 7.6).

The fact that moments of intensity behave in the layer as exponentially increasing functions is evidence of the *phenomenon of stochastic wave parametric resonance*, which is similar to the ordinary parametric resonance. The only difference consists in the fact that values of intensity moments at layer boundary are asymptotically predetermined; as a result, the wavefield intensity exponentially increases inside the layer and its maximum occurs approximately in the middle of the layer.

In the limit of a half-space ($L_0 \to -\infty$), the region of the exponential growth of all moments beginning from the second one occupies the whole of the half-space, and $\left\langle \overline{I(x, L)} \right\rangle = 2$.

Now, we turn back to the equation for moments of opposite wave intensities in non-absorptive medium, i.e., to Eq. (7.29) at $\beta = 0$ in the limit of the half-space $(L_0 \rightarrow -\infty)$ filled with random medium. In this case, $W_L = 1$ with probability one, and the solution to Eq. (7.29) has the form

$$Q(x, L; \mu, \lambda, W) = \delta(W - 1)e^{D\lambda(\lambda-1)(L-x)},$$

so that

$$\left\langle W_1^{\lambda-\mu}(x, L)W_2^{\mu}(x, L)\right\rangle = e^{D\lambda(\lambda-1)(L-x)}. \tag{7.44}$$

In view of arbitrariness of parameters λ and μ, this means that

$$W_1(x, L) = W_2(x, L) = W(x, L)$$

with probability one and quantity $W(x, L)$ has the lognormal probability density. In addition, the mean value of this quantity is equal to unity, and its higher moments beginning from the second one exponentially increase with the distance in the medium

$$\langle W(x, L)\rangle = 1, \quad \left\langle W^n(x, L)\right\rangle = e^{Dn(n-1)(L-x)}, \quad n = 2, 3, \cdots. \tag{7.45}$$

Note that wavefield intensity $I(x, L)$ has in this case the form

$$I(x, L) = 2W(x, L)(1 + \cos\phi_x), \tag{7.46}$$

where ϕ_L is the phase of the reflection coefficient.

In accordance with the properties of lognormal distribution, the typical realization curve of function $W(x, L)$ is the curve exponentially decaying with distance in the medium

$$W^*(x, L) = e^{-D(L-x)}, \tag{7.47}$$

and this function is related to the *Lyapunov exponent*.

In addition, realizations of function $W(x, L)$ satisfy the inequality

$$W(x, L) < 4e^{-D(L-x)/2}$$

within the whole of the half-space with probability equal to 1/2.

In physics of disordered systems, the exponential decay of typical realization curve (7.47) with increasing $\xi = D(L - x)$ is usually identified with the property of *dynamic localization*, and quantity

$$l_{\text{loc}} = \frac{1}{D}$$

is usually called the *localization length*. Here,

$$l_{\text{loc}}^{-1} = -\frac{\partial}{\partial L} \langle \varkappa(x, L) \rangle,$$

where

$$\varkappa(x, L) = \ln W(x, L).$$

Physically, the lognormal property of wavefield intensity $W(x, L)$ implies the existence of large spikes relative typical realization curve (7.47) towards both large and small intensities. This result agrees with the example of simulations given in Sect. 3.2 (see Fig. 3.5). However, these spikes of intensity contain only small energy, because random area

$$S(L) = D \int\limits_{-\infty}^{L} dx \, W(x, L),$$

below curve $W(x, L)$, has, in accordance with the lognormal probability distribution attribute, the steady-state (independent of L) probability density

$$P(S) = \frac{1}{S^2} \exp\left\{-\frac{1}{S}\right\}$$

that coincides with the distribution of total energy of the wavefield in the half-space (7.18) if we set $E = 2S$. This means that the term dependent on fast phase oscillations of reflection coefficient in Eq. (7.46) only slightly contributes to total energy.

Thus, the knowledge of the one-point probability density provides an insight into the evolution of separate realizations of wavefield intensity in the whole space and allows estimating the parameters of this evolution in terms of statistical characteristics of fluctuating medium.

7.2.3   Dissipative Medium

In the presence of a finite (even arbitrary small) absorption in the medium occupying the half-space, the exponential growth of moment functions must cease and give place to attenuation.

If $\beta \gg 1$ (i.e., if the effect of absorption is dominated by the effect of diffusion), then

$$P(W) = 2\beta e^{-2\beta W},$$

and, as can be easily seen from Eq. (7.29), opposite wave intensities $W_1(x; L)$ and $W_2(x; L)$ appear statistically independent, i.e., uncorrelated. In this case,

$$\langle W_1(\xi) \rangle = \exp\left\{-\beta\xi\left(1 + \frac{1}{\beta}\right)\right\}, \quad \langle W_2(\xi) \rangle = \frac{1}{2\beta} \exp\left\{-\beta\xi\left(1 + \frac{1}{\beta}\right)\right\}.$$

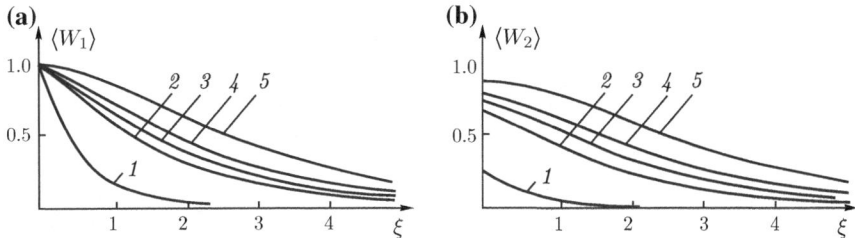

Fig. 7.7 Distribution of wavefield average intensity along the medium; **a** the transmitted wave and **b** the reflected wave. *Curves 1–5* correspond to parameter $\beta = 1, 0.1, 0.06, 0.04$ and 0.02, respectively

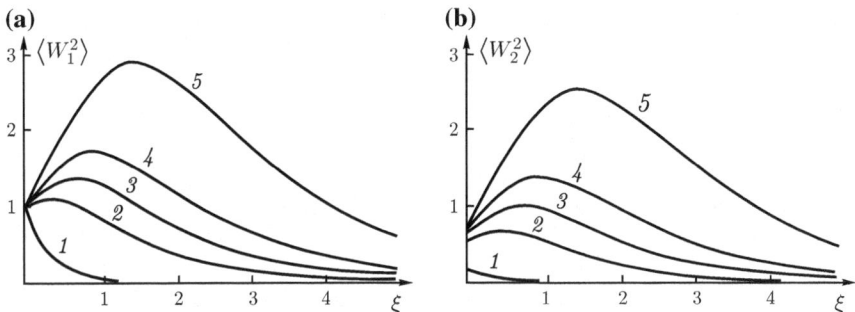

Fig. 7.8 Distribution of the second moment of wavefield intensity along the medium; **a** the transmitted wave and **b** the reflected wave. *Curves 1–5* correspond to parameter $\beta = 1, 0.1, 0.06, 0.04$ and 0.02, respectively

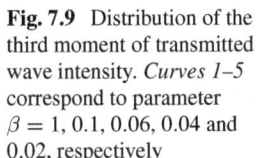

Fig. 7.9 Distribution of the third moment of transmitted wave intensity. *Curves 1–5* correspond to parameter $\beta = 1, 0.1, 0.06, 0.04$ and 0.02, respectively

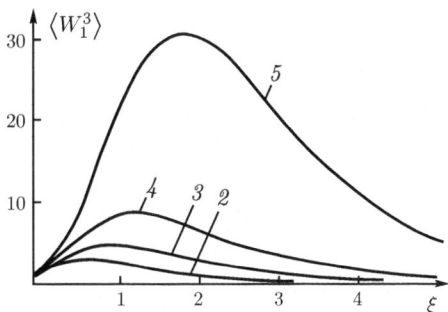

Figures 7.7, 7.8, 7.9 and 7.10 show the examples of moment functions of random processes obtained by numerical solution of Eq. (7.29) and calculation of quadrature (7.30) for different values of parameter β. Different figures mark the curves corresponding to different values of parameter β. Figure 7.7 shows average intensities of the transmitted and reflected waves. The curves monotonically decrease with increasing ξ. Figure 7.8 shows the corresponding curves for second moments. We see that $\langle W_1^2(0)\rangle = 1$ and $\langle W_2^2(0)\rangle = \langle |R_L|^4\rangle$ at $\xi = 0$

Fig. 7.10 Correlation between the intensities of transmitted and reflected waves. *Curves 1–5* correspond to parameter $\beta = 1, 0.1, 0.06, 0.04$ and 0.02, respectively

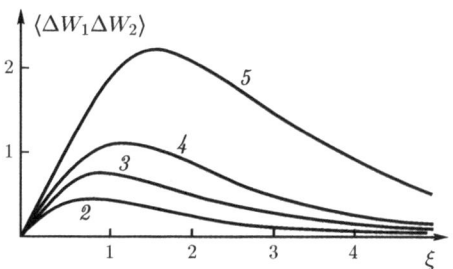

$$\langle W_1^2(0)\rangle = 1, \quad \langle W_2^2(0)\rangle = \langle |R_L|^4\rangle.$$

For $\beta < 1$, the curves as functions of ξ become nonmonotonic; the moments first increase, then pass the maximum, and finally monotonically decay. With decreasing parameter β, the position of the maximum moves to the right and the maximum value increases. Figure 7.9 shows the similar curves for the third moment $\langle W_1^3(\xi)\rangle$, and Fig. 7.10 shows curves for mutual correlation of intensities of the transmitted and reflected waves $\langle \Delta W_1(\xi)\Delta W_2(\xi)\rangle$ (here, $\Delta W_n(\xi) = W_n(\xi) - \langle W_n(\xi)\rangle$). For $\beta \geq 1$, this correlation disappear. For $\beta < 1$, the correlation is strong, and wave division into opposite waves appears physically senseless, but mathematically useful technique. For $\beta \geq 1$, such a division is justified in view of the lack of mutual correlation.

The fact that moments of intensity behave in the layer as exponentially increasing functions is evidence of the *phenomenon of stochastic wave parametric resonance*, which is similar to the ordinary parametric resonance. The only difference consists in the fact that values of intensity moments at layer boundary are asymptotically predetermined; as a result, the wavefield intensity moments exponentially increase inside the layer forming maxima in the middle of the layer.

As was shown earlier, in the case of the half-space of random medium with $\beta = 0$, all wavefield moments beginning from the second one exponentially increase with the distance the wave travels in the medium. It is clear, that problem solution for small β ($\beta \ll 1$) must show the singular behavior in β in order to vanish the solution for sufficiently long distances. Consider this asymptotic case in more detail.

We introduce function

$$Q(x, L; \mu, \lambda, u) = \left\langle W_1^{\mu-\lambda}(x, L)W_2^{\lambda}(x, L)\delta(u_L - u)\right\rangle$$

satisfying in the case of a half-space the equation

$$\frac{\partial}{\partial \xi} Q(\xi; \mu, \lambda, u) = \left(-\beta\mu + \beta\frac{\partial}{\partial u}\left(u^2 - 1\right) + \mu(\mu + 1) - \frac{2\mu^2}{u+1}\right) Q(\xi; \mu, \lambda, u)$$

$$+ \left[2\mu(u - 1)\frac{\partial}{\partial u} + \frac{\partial}{\partial u}\left(u^2 - 1\right)\frac{\partial}{\partial u}\right] Q(\xi; \mu, \lambda, u) \qquad (7.48)$$

with the initial value

$$Q(0; \mu, \lambda, u) = \left(\frac{u-1}{u+1}\right)^{\lambda} P(u),$$

where $\xi = D(L - x) > 0$ and $P(u)$ is the steady-state probability density (7.12).

Our interest is in quantities

$$\left\langle W_1^{\mu-\lambda}(\xi) W_2^{\lambda}(\xi) \right\rangle = \int_1^{\infty} du \, Q(\xi; \mu, \lambda, u).$$

Replace variable $u \to \beta(u - 1)$ in Eq. (7.48) and perform limit process $\beta \to 0$. As a result, we obtain a simpler equation

$$\frac{\partial}{\partial \xi} Q(\xi; \mu, \lambda, u) = \left(\mu(\mu + 1) - \frac{2\mu^2 \beta}{u} + 2\mu u \frac{\partial}{\partial u} + \frac{\partial}{\partial u} u^2 \frac{\partial}{\partial u} \right) Q(\xi; \mu, \lambda, u),$$

(7.49)

with the initial value $Q(0; \mu, \lambda, u) = e^{-u}$.

The solution to this equation as a function of variable u (and, consequently, parameter β) has a singularity in the case of arbitrary small, but finite absorption in the medium. This solution can be obtained using the integral Kontorovich–Lebedev transform. As a result, in the case of integer parameters $\mu = n$, $\lambda = m$, we obtain the asymptotic representation in the form of the quadrature

$$\left\langle W_1^{n-m}(\xi) W_2^m(\xi) \right\rangle = \frac{4}{\pi (\varepsilon n)^{2n-1}} \int_0^{\infty} d\tau \, \tau \sinh\left(\frac{\pi\tau}{2}\right) e^{-\xi(1+\tau^2)/4} g_n(\tau) \psi_0(\tau),$$

$$\psi_0(\tau) = \int_0^{\infty} \frac{dy}{y^{2(n+1)}} \frac{e^{-1/y^2}}{\left(1 + 2\beta y^2\right)^m} K_{i\tau}(\varepsilon n y),$$

(7.50)

where $\varepsilon = \sqrt{8\beta}$, $g_n(\tau) = \left[(2n-3)^2 + \tau^2\right] g_{n-1}(\tau)$, $g_1(\tau) = 1$, and $K_{i\tau}(x)$ is the *imaginary index McDonalds function* of the first kind satisfying equations

$$\left(x^2 \frac{d^2}{dx^2} + x \frac{d}{dx} - x^2 + \tau^2 \right) K_{i\tau}(x) = 0,$$

(7.51)

$$\left(\frac{d}{dx} x^2 \frac{d}{dx} - x \frac{d}{dx} \right) K_{i\tau}(x) = \left(x^2 - \tau^2 \right) K_{i\tau}(x).$$

From Eq. (7.50), we see that, in asymptotic limit $\beta \ll 1$, intensities of opposite waves are equal with probability one, and the solution for small distances from the boundary coincides with the solution corresponding to the stochastic parametric resonance.

For sufficiently great distances ξ, namely

$$\xi \gg 4 \left(n - \frac{1}{2} \right) \ln \left(\frac{n}{\beta} \right),$$

quantities $\langle W^n(\xi) \rangle$ are characterized by the universal spatial localization behavior

$$\langle W^n(\xi) \rangle \cong A_n \frac{1}{\beta^{n-1/2}} \ln \left(\frac{1}{\beta} \right) \frac{1}{\xi \sqrt{\xi}} e^{-\xi/4},$$

which coincides, to a numerical factor, with the asymptotic behavior of moments of the transmission coefficient of a wave passed through the layer of thickness ξ in the case $\beta = 0$.

Thus, the behavior of moments of opposite wave intensities appears essentially different in three regions. In the first region (it corresponds to the stochastic parametric resonance), the moments exponentially increase with the distance in medium and wave absorption plays only insignificant role. In the second region, absorption plays the most important role, because namely absorption ceases the exponential growth of moments. In the third region, the decrease of moment functions of opposite wave intensities is independent of absorption. The boundaries of these regions depend on parameter β and tend to infinity for $\beta \to 0$.

Note that, in the general case of arbitrary parameter β, mean logarithm of forward wave and its variance are given, in accordance with Eq. (7.23), by the relationships

$$\langle \varkappa_1(x, L) \rangle = -(1 + \beta)\,\xi, \quad \sigma^2_{\varkappa_1}(x, L) = 2 \langle |R_L|^2 \rangle \xi, \tag{7.52}$$

where $\langle |R_L|^2 \rangle$ is given by Eq. (7.13).

7.2.4 Plane Wave Source Located in Random Medium

Consider now the asymptotic solution of the problem on the plane wave source in infinite space ($L_0 \to -\infty$, $L \to \infty$) under the condition $\beta \to 0$. In this case, it appears convenient to calculate average wavefield intensity in region $x < x_0$ using relationships (3.58) and (3.36)

$$\beta \langle I(x; x_0) \rangle = \frac{1}{D} \frac{\partial}{\partial x} \langle S(x; x_0) \rangle = \frac{1}{D} \frac{\partial}{\partial x} \langle \psi(x; x_0) \rangle,$$

where quantity

$$\psi(x; x_0) = \exp \left\{ -\beta D \int_x^{x_0} d\xi \frac{|1 + R_\xi|^2}{1 - |R_\xi|^2} \right\}$$

satisfies, as a function of parameter x_0, the stochastic equation

$$\frac{\partial}{\partial x_0}\psi(x; x_0) = -\beta D\frac{|1 + R_{x_0}|^2}{1 - |R_{x_0}|^2}\psi(x; x_0), \quad \psi(x; x) = 1.$$

Introduce function

$$\varphi(x; x_0; u) = \psi(x; x_0)\delta(u_{x_0} - u), \tag{7.53}$$

where function $u_L = (1 + W_L)/(1 - W_L)$ satisfies the stochastic system of equations
(7.5). Differentiating Eq. (7.53) with respect to x_0, we obtain the stochastic equation

$$\frac{\partial}{\partial x_0}\varphi(x; x_0; u) = -\beta D\left\{u + \sqrt{u^2 - 1}\cos\phi_{x_0}\right\}\varphi(x; x_0; u)$$

$$+ \beta D\frac{\partial}{\partial u}\left\{\left(u^2 - 1\right)\varphi(x; x_0; u)\right\}$$

$$- k\varepsilon_1(x_0)\frac{\partial}{\partial u}\left\{\sqrt{u^2 - 1}\sin\phi_{x_0}\varphi(x; x_0; u)\right\}. \tag{7.54}$$

Average now Eq. (7.54) over an ensemble of realizations of random process $\varepsilon_1(x_0)$
assuming it, as earlier, the Gaussian process delta-correlated in x_0. After additionally
averaging over fast oscillations (over the phase of reflection coefficient), we obtain
that function

$$\Phi(\xi; u) = \langle\varphi(x; x_0; u)\rangle = \langle\psi(x; x_0)\delta(u_{x_0} - u)\rangle,$$

where $\xi = D|x - x_0|$, satisfies the equation

$$\frac{\partial}{\partial\xi}\Phi(\xi; u) = -\beta u\Phi(\xi; u) + \beta\frac{\partial}{\partial u}\left(u^2 - 1\right)\Phi(\xi; u)$$

$$+ \frac{\partial}{\partial u}\left(u^2 - 1\right)\frac{\partial}{\partial u}\Phi(\xi; u) \tag{7.55}$$

with the initial condition

$$\Phi(0; u) = P(u) = \beta e^{-\beta(u-1)}.$$

The average intensity can now be represented in the form

$$\beta\langle I(x; x_0)\rangle = -\frac{\partial}{\partial\xi}\int_1^\infty du\Phi(\xi; u) = \beta\int_1^\infty du u\Phi(\xi; u).$$

Equation (7.55) allows limiting process $\beta \to 0$. As a result, we obtain a simpler
equation

$$\frac{\partial}{\partial \xi} \widetilde{\Phi}(\xi; u) = -u \widetilde{\Phi}(\xi; u) + \frac{\partial}{\partial u} u^2 \widetilde{\Phi}(\xi; u) + \frac{\partial}{\partial u} u^2 \frac{\partial}{\partial u} \widetilde{\Phi}(\xi; u),$$

(7.56)

$$\widetilde{\Phi}(0; u) = e^{-u}.$$

Consequently, localization of average intensity in space is described by the quadrature

$$\Phi_{\text{loc}}(\xi) = \int_1^\infty du\, u \widetilde{\Phi}(\xi; u),$$

where

$$\Phi_{\text{loc}}(\xi) = \lim_{\beta \to 0} \beta \left\langle I(x; x_0) \right\rangle = \lim_{\beta \to 0} \frac{\left\langle I(x; x_0) \right\rangle}{\left\langle I(x_0; x_0) \right\rangle}.$$

Thus, the average intensity of the wavefield generated by the point source has for $\beta \ll 1$ the following asymptotic behavior

$$\left\langle I(x; x_0) \right\rangle = \frac{1}{\beta} \Phi_{\text{loc}}(\xi).$$

Equation (7.56) can be easily solved with the use of the Kontorovich–Lebedev transform; as a result, we obtain the expression for the *localization curve*

$$\Phi_{\text{loc}}(\xi) = 2\pi \int_0^\infty d\tau\, \tau \left(\tau^2 + \frac{1}{4} \right) \frac{\sinh(\pi\tau)}{\cosh^2(\pi\tau)} e^{-\left(\tau^2 + \frac{1}{4}\right)\xi}.$$

(7.57)

For small distances ξ, the localization curve decays according to relatively fast law

$$\Phi_{\text{loc}}(\xi) \approx e^{-2\xi}.$$

(7.58)

For great distances ξ (namely, for $\xi \gg \pi^2$), it decays significantly slower, according to the universal law

$$\Phi_{\text{loc}}(\xi) \approx \frac{\pi^2 \sqrt{\pi}}{8} \frac{1}{\xi \sqrt{\xi}} e^{-\xi/4},$$

(7.59)

but for all that

$$\int_0^\infty d\xi \Phi_{\text{loc}}(\xi) = 1.$$

Function (7.57) is given in Fig. 7.11, where asymptotic curves (7.58) and (7.59) are also shown for comparison purposes.

Fig. 7.11 Localization
curve for a source in infinite
space (7.57) (*curve a*).
Curves b and *c* correspond to
asymptotic expressions for
small and large distances
from the source

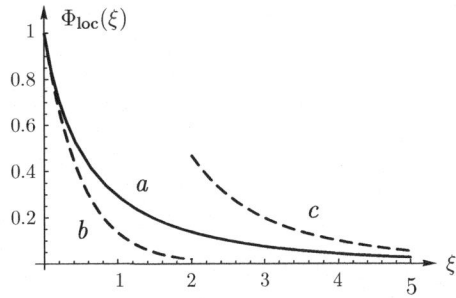

A similar situation occurs in the case of the plane wave source located at the reflecting boundary.

In this case, we obtain the expression

$$\lim_{\beta \to 0} \frac{\langle I_{\text{ref}}(x; L) \rangle}{\langle I_{\text{ref}}(L; L) \rangle} = \frac{1}{2} \Phi_{\text{loc}}(\xi), \quad \xi = D(L - x). \tag{7.60}$$

7.3 Numerical Simulation

The above theory rests on two simplifications—on using the delta-correlated approximation of function $\varepsilon_1(x)$ (or the diffusion approximation) and finding slow (within the scale of a wavelength) variations of statistical characteristics by averaging over fast oscillations. Averaging over fast oscillations is validated only for statistical characteristics of the reflection coefficient in the case of random medium occupying a half-space. For statistical characteristics of the wavefield intensity in medium, the corresponding validation appears very difficult if at all possible (this method is merely physical than mathematical). Numerical simulation of the exact problem offers a possibility of both verifying these simplifications and obtaining the results concerning more difficult situations for which no analytic results exists.

In principle, such numerical simulation could be performed by way of multiply solving the problem for different realizations of medium parameters followed by averaging the obtained solutions over an ensemble of realizations. However, such an approach is not very practicable because it requires a vast body of realizations of medium parameters. Moreover, it is unsuitable for real physical problems, such as wave propagation in Earth's atmosphere and ocean, where only a single realization is usually available. A more practicable approach is based on the ergodic property of boundary-value problem solutions with respect to the displacement of the problem along the single realization of function $\varepsilon_1(x)$ defined along the half-axis (L_0, ∞) (see Fig. 7.12).

This approach assumes that statistical characteristics are calculated by the formula

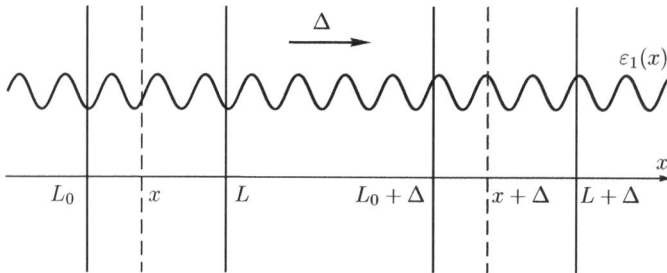

Fig. 7.12 Averaging over parameter Δ by the procedure based on ergodicity of imbedding equations for a half-space of random medium

$$\langle F(L_0; x, x_0; L) \rangle = \lim_{\delta \to \infty} F_\delta(L_0; x, x_0; L),$$

where

$$F_\delta(L_0; x, x_0; L) = \frac{1}{\delta} \int_0^\delta d\Delta F(L_0 + \Delta; x + \Delta, x_0 + \Delta; L + \Delta).$$

In the limit of a half-space ($L_0 \to -\infty$), statistical characteristics are independent of L_0, and, consequently, the problem have ergodic property with respect to the position of the right-hand layer boundary L (simultaneously, parameter L is the variable of the imbedding method, because this position is identified in this case with the displacement parameter). As a result, having solved the imbedding equation for the sole realization of medium parameters, we simultaneously obtain all desired statistical characteristics of this solution by using the obvious formula

$$\langle F(x, x_0; L) \rangle = \frac{1}{\delta} \int_0^\delta d\xi F\left(\xi, \xi + x_0 - x; \xi + (L - x_0) + (x_0 - x)\right)$$

for sufficiently large interval $(0, \delta)$. This approach offers a possibility of calculating even the wave statistical characteristics that cannot be obtained within the framework of current statistical theory, and this calculation requires no additional simplifications.

In the case of the layer of finite thickness, the problem is not ergodic with respect to parameter L. However, the corresponding solution can be expressed in terms of two independent solutions of the problem on the half-space [57] and, consequently, it can be reduced to the case ergodic with respect to L.

Systematically, the program of numerical simulation was implemented in paper [58] (see also monograph [29]).

Consider several particular results obtained with numerical simulation.

7.3.1 Wave Incident on the Medium Layer

The first stage of simulations consisted in studying the moments of the reflection coefficient. Figure 7.13 shows the modulus of reflection coefficient correlation function. The solid line corresponds to ensemble averaging, the circles (∘) correspond to averaging over the realization of length $L = 10$, and dots (•) correspond to averaging over the realization of length $L = 300$.

The second stage of simulations consisted in studying the first and second moments of the wavefield intensity $I(x, L)$ in the problem on the wave incident on the gyrandom half-space. Simultaneously, we investigated the dependence of the result on the length of sampling used for averaging. Simulated results were compared with the above theoretical results.

Figure 7.14 shows moments of the intensity simulated with $\beta = 1$. Curves *1* and *2* show average intensity $\langle I(x, L) \rangle$ and average square of the intensity $\langle I^2(x, L) \rangle$ calculated with the use of ensemble averaging. The calculation showed that samplings of dimensionless length $L \sim 10 - 20$ are sufficient for obtaining satisfactory results. For $\beta = 0.08$, such a sampling appears insufficient, and obtaining the

Fig. 7.13 Modulus of reflection coefficient correlation function $\left| \left\langle R_h R_{h+\xi}^* \right\rangle \right|$ at $\beta = 0.08$ as a function of parameter ξ

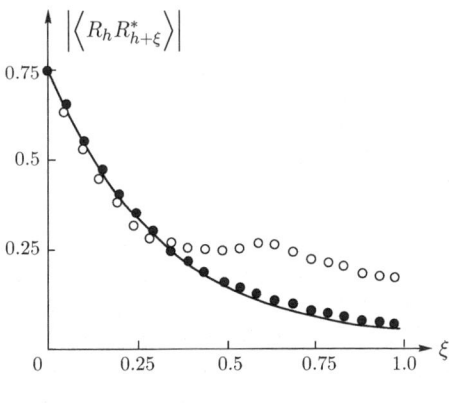

Fig. 7.14 Moments of wavefield intensity in the problem on a wave incident on medium layer ($\beta = 1$)

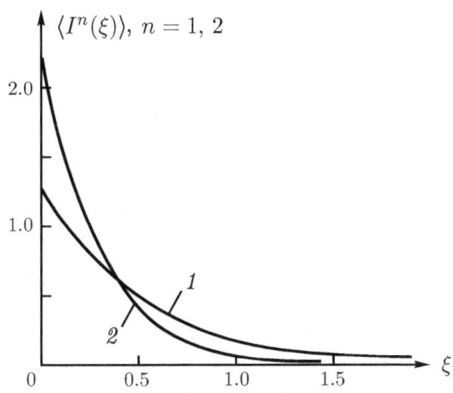

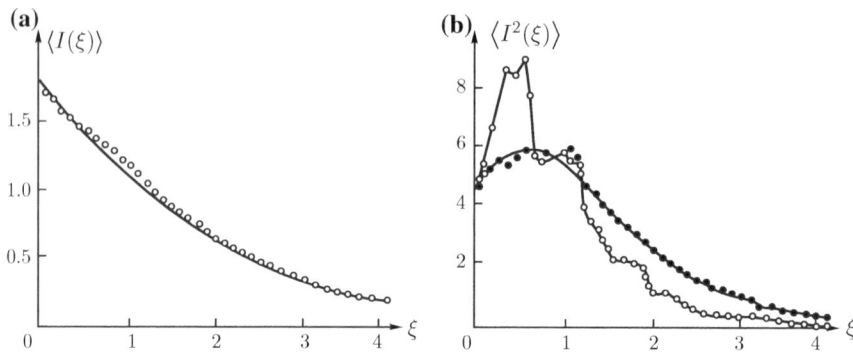

Fig. 7.15 Moments of wavefield intensity in the problem on a wave incident on medium layer ($\beta = 0.08$). **a** Average intensity $\langle I(x, L)\rangle$ and **b** average square of the intensity $\langle I^2(x, L)\rangle$

Fig. 7.16 Average intensity $\langle I(x, x_0)\rangle$ of the field generated by a source in infinite space

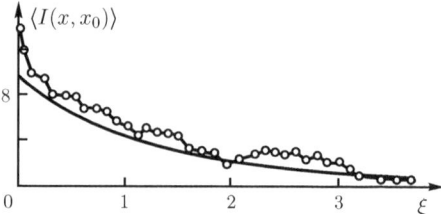

adequate result requires samplings of length $L \sim 300$ (Fig. 7.15). The solid lines correspond to ensemble averaging, circles ($\circ$) correspond to averaging over a realization of length $L = 10$, and dots ($\bullet$) correspond to averaging over a realization of length $L = 300$.

7.3.2 Plane Wave Source in the Medium Layer

Figure 7.16 shows moment $\langle I(x, x_0)\rangle$ simulated in the case of the source in infinite space for sampling length $L = 10$ and $\beta = 0.08$. The solid line corresponds to ensemble averaging and circles ($\circ$) correspond to averaging over a realization of length $L = 10$. We can see from this pattern that even such short sampling adequately catches the behavioral tendency of average intensity of the field generated by a source in infinite space. All other curves were obtained with sampling length $L = 300 - 400$.

Figure 7.17 shows average intensity of the field generated by a source simulated with $\beta = 1$ for different boundary conditions. Again, the solid lines correspond to ensemble averaging, circles ($\circ$) correspond to simulations for free passage through the boundary, dots ($\bullet$) correspond to simulations for reflecting boundary with the condition $dG(H; x_0)/dx = 0$, and crosses ($\times$) correspond to simulations for reflecting

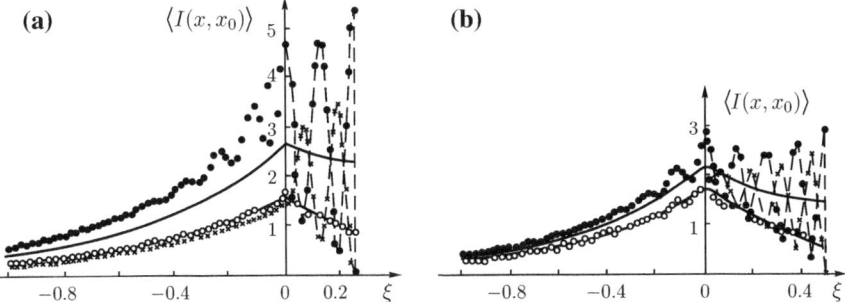

Fig. 7.17 Average intensity of the source-generated field for $\beta = 1$ and boundary positions **a** $DH = 0.25$ and **b** $DH = 0.5$

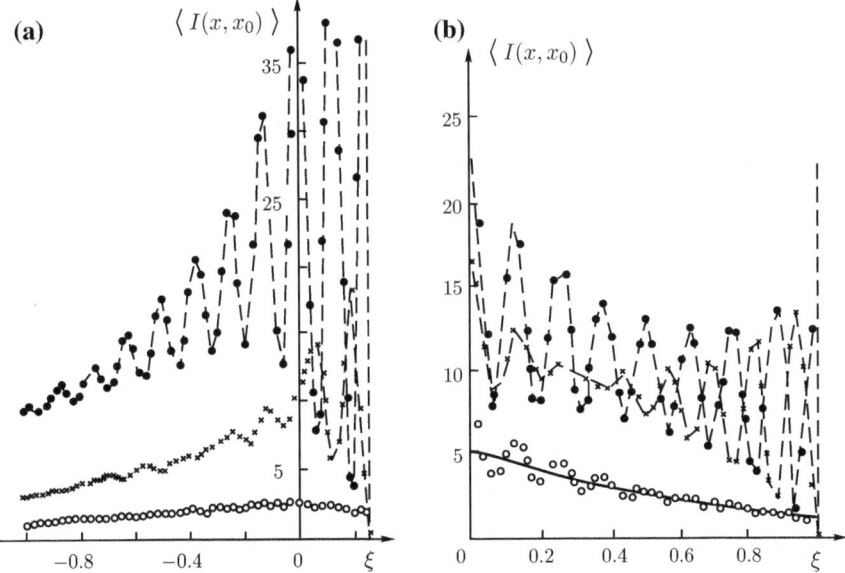

Fig. 7.18 Average intensity of the source-generated field for $\beta = 0.08$ and boundary positions **a** $DH = 0.25$ and **b** $DH = 1$

boundary with the condition $G(H; x_0) = 0$. Figure 7.17 shows that simulated results are in adequate agreement with theoretical curves in the case of the penetrating boundary; at the same time, it shows that, in the case of reflecting boundary, average intensity is strongly oscillating, which indicates that the interference pattern of average intensity appears to be complicated even at $\beta = 1$. The amplitude of oscillations decreases with moving the source away from the boundary. Figure 7.18 shows similar curves simulated with $\beta = 0.08$. This case is characterized by more intense

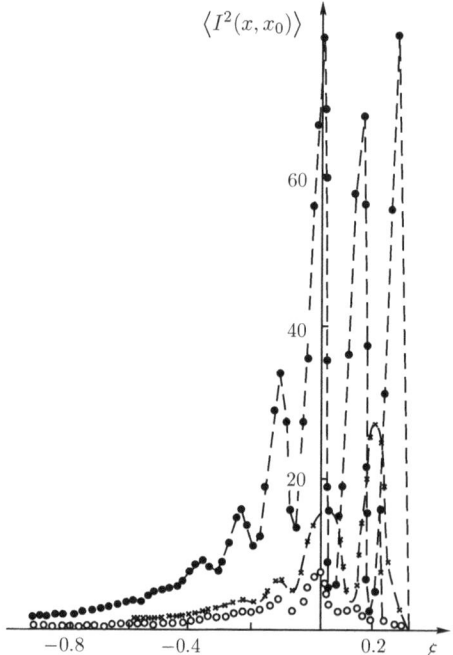

Fig. 7.19 Second moment of intensity of the source-generated field for $\beta = 1$ and boundary position $DH = 0.25$

variations of function $\langle I(x, x_0)\rangle$. Again, the amplitude of oscillations decreases with moving the source away from the boundary.

The method of numerical simulations enables us to find the statistical characteristics that cannot be determined theoretically yet. Figures 7.19 and 7.20 show the simulated second moments of intensity of the field generated by a source $\langle I^2(x, x_0)\rangle$ for $\beta = 1$ and $\beta = 0.08$ and different boundary conditions. Circles (o) correspond to simulations for free passage through the boundary, dots (•) correspond to simulations for reflecting boundary with the condition $dG(H; x_0)/dx = 0$, and crosses (×) correspond to simulations for reflecting boundary with the condition $G(H; x_0) = 0$. The second moments oscillate with the same period, but oscillating amplitude significantly increases.

As can be seen from the above figures, the oscillations of period ~ 0.13 are characteristic of the moments of wavefield intensity in the presence of boundary. These oscillations are related to our choice of wave parameter $\alpha = 25$, because the corresponding period $T = \pi/\alpha = 0.126$.

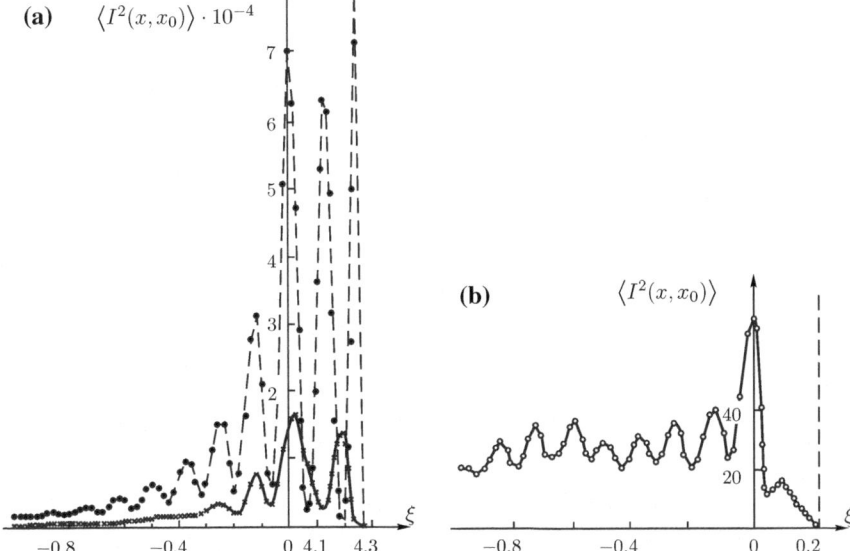

Fig. 7.20 Second moment of intensity of the source-generated field for $\beta = 0.08$ and **a** reflecting boundary at $DH = 4.3$ and **b** freely penetrating boundary at $DH = 0.25$

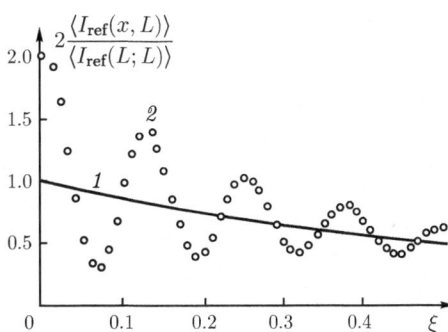

Fig. 7.21 *Curve 1* shows the localization curve (7.57) and *circles 2* show the simulated result

Figure 7.21 shows the mean intensity $2 \langle I_{\text{ref}}(x, L) \rangle / \langle I_{\text{ref}}(L; L) \rangle$ of the field of a source located at reflecting boundary $x_0 = L$ with boundary condition $dG(x, x_0; L)/dx|_{x=L} = 0$ for parameters $\beta = 0.08$ and $k/D = 25$. In region $\xi = D(L - x) < 0.3$, one can see oscillations of period $T = 0.13$. For larger ξ, simulated results agree well with localization curve (7.60).

Chapter 8
Lognormal Fields, Statistical Topography, and Clustering

8.1 Lognormal Random Fields

Let us consider now a positive lognormal random field $f(\mathbf{r}, t)$, whose one-point probability density

$$P(\mathbf{r}, t; f) = \langle \delta(f(\mathbf{r}, t) - f) \rangle$$

is governed by the equation

$$\frac{\partial}{\partial t} P(\mathbf{r}, t; f) = \left\{ D_0 \frac{\partial^2}{\partial \mathbf{r}^2} + \alpha \frac{\partial}{\partial f} f + D_f \frac{\partial}{\partial f} f \frac{\partial}{\partial f} f \right\} P(\mathbf{r}, t; f), \qquad (8.1)$$

with the initial condition $P(\mathbf{r}, 0; f) = \delta(f - f_0(\mathbf{r}))$, where D_0 is the diffusion coefficient in the $\mathbf{r}$-space, and the coefficients α and D_f characterize diffusion in the f-space. The parameter α can both differ from zero and be equal to it (the critical case). The change in the sign of α for one-point characteristics implies the transition from the field $f(\mathbf{r}, t)$ to the field $\tilde{f}(\mathbf{r}, t) = 1/f(\mathbf{r}, t)$.

The solution of Eq. (8.1) is written out as follows:

$$P(\mathbf{r}, t; f) = \frac{1}{2f\sqrt{\pi D_f t}} \exp\left\{ D_0 t \frac{\partial^2}{\partial \mathbf{r}^2} \right\} \exp\left\{ -\frac{\ln^2\left[f e^{\alpha t}/f_0(\mathbf{r}) \right]}{4 D_f t} \right\}. \qquad (8.2)$$

Note that for a positive conservative random field $f(\mathbf{r}, t)$, satisfying the condition $\int d\mathbf{r}\, f(\mathbf{r}, t) = \int d\mathbf{r}\, f_0(\mathbf{r})$, the parameter $\alpha = D_f$ and Eq. (8.1) can be recast as

$$\frac{\partial}{\partial t} P(\mathbf{r}, t; f) = \left(D_0 \frac{\partial^2}{\partial \mathbf{r}^2} + \alpha \frac{\partial^2}{\partial f^2} f^2 \right) P(\mathbf{r}, t; f). \qquad (8.3)$$

© Springer International Publishing AG 2017
V.I. Klyatskin, *Fundamentals of Stochastic Nature Sciences*,
Understanding Complex Systems, DOI 10.1007/978-3-319-56922-2_8

Needless to say, the property of intermittency is always exhibited by any random field $f(\mathbf{r}, t)$ as well. For any fixed point in space $\mathbf{r}$, the temporal evolution of $f(\mathbf{r}, t)$ is a random process for which all the above holds.

For a spatially, statistically homogeneous problem which corresponds to the initial field distribution $f_0(\mathbf{r}) = f_0$, no one-point statistical characteristics of the field $f(\mathbf{r}, t)$ depend on the point $\mathbf{r}$, and the positivity of the index $\alpha = -\lim\limits_{t \to \infty} \dfrac{\partial \langle \ln f(\mathbf{r}, t) \rangle}{\partial t}$ for the lognormal field $f(\mathbf{r}, t)$ implies that, at any location in space, the realizations of this field decay with time despite large rare excursions, which occur for a lognormal process. In this case, the characteristic time of field decay is $t \sim 1/\alpha$. *But if the field decays almost everywhere, it must be concentrated somewhere, i.e., clustering should take place.* For a negative parameter α the field grows at every fixed point in space.

In the last case, probability density (8.2) does not depend on $\mathbf{r}$ and is described by the equation

$$\frac{\partial}{\partial t} P(t; f) = \left\{ \alpha \frac{\partial}{\partial f} f + D_f \frac{\partial}{\partial f} f \frac{\partial}{\partial f} f \right\} P(t; f),$$
$$P(0; f) = \delta(f - f_0), \tag{8.4}$$

with the solution

$$P(t; f) = \frac{1}{2f\sqrt{\pi D_f t}} \exp\left\{ -\frac{\ln^2\left[f e^{\alpha t}/f_0\right]}{4 D_f t} \right\}. \tag{8.5}$$

Thus, for a spatially homogeneous problem statement, the one-point statistical characteristics of a random field $f(\mathbf{r}, t)$ are statistically equivalent to those of the lognormal process $f(t; \alpha)$ with the probability density (8.5). A specific feature of this distribution is the appearance of a long gently sloping *'tail'* for $D_f t \gg 1$, which indicates the increased role of large fluctuations of the process $f(t; \alpha)$ in forming one-time statistics (see Fig. 5.1). For this distribution, all moment functions exponentially grow with time and, in particular, at $n = 1$ and for $D_f > \alpha$ the expectation is given by

$$\langle f(\mathbf{r}, t) \rangle = f_0 e^{(D_f - \alpha) t},$$

whereas the quantity α is the Lyapunov characteristic index.

Figure 8.1 plots schematically random realizations of the field $f(\mathbf{r}, t)$ for the parameter α of different signs.

One can describe spatial clustering in almost any realization of the random field $f(\mathbf{r}, t)$ by resorting to the ideas of statistical topography.

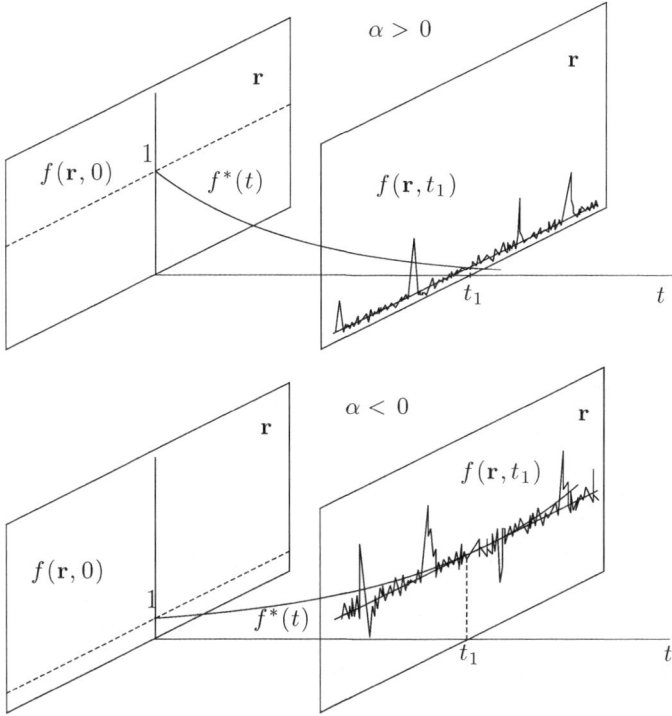

Fig. 8.1 Schematics of the behaviour of random realizations of the field $f(\mathbf{r}, t)$ for $\alpha > 0$ and for $\alpha < 0$

8.2 Statistical Topography of Random Fields

The main subject of study in the statistical topography of random fields, just as in the traditional topography of massifs, is a system of contours—isolines (in the two-dimensional case) or isosurfaces (in three dimensions) defined by the equality $f(\mathbf{r}, t) = f = \text{const}$.

To analyze the system of contours (we limit ourselves to the two-dimensional case $\mathbf{r} = \mathbf{R}$) for simplicity), it is convenient to introduce the Dirac delta-function constrained to these contours:

$$\varphi(\mathbf{R}, t; f) = \delta(f(\mathbf{R}, t) - f), \tag{8.6}$$

called the *indicator function*.

Such quantities as the total area bounded by the level lines of the areas where the random field $f(\mathbf{R}, t)$ exceeds the given level, i.e., $f(\mathbf{R}, t) > f$:

$$S(t; f) = \int d\mathbf{R}\, \theta(f(\mathbf{R}, t) - f) = \int d\mathbf{R} \int\limits_{f}^{\infty} df'\, \varphi(\mathbf{R}, t; f'),$$

and the total '*mass*' of the field comprised in these regions, namely

$$M(t; f) = \int d\mathbf{R}\, f(\mathbf{R}, t)\theta(f(\mathbf{R}, t) - f) = \int d\mathbf{R} \int\limits_{f}^{\infty} df'\, f'\varphi(\mathbf{R}, t; f'),$$

where $\theta(f(\mathbf{R}, t) - f)$ is the Heaviside theta-function, may be expressed in terms of function (8.6).

The mean value of the indicator function (8.6) over an ensemble of realizations of the random field $f(\mathbf{R}, t)$ defines the one-time and one-point in space probability density [28, 29]

$$P(\mathbf{R}, t; f) = \langle \delta(f(\mathbf{R}, t) - f) \rangle,$$

which directly sets the ensemble mean values of quantities $S(t; f)$ and $M(t; f)$:

$$\langle S(t; f) \rangle = \int d\mathbf{R} \int\limits_{f}^{\infty} df'\, P(\mathbf{R}, t; f'), \quad \langle M(t; f) \rangle = \int d\mathbf{R} \int\limits_{f}^{\infty} df'\, f' P(\mathbf{R}, t; f').$$

Information on the detailed structure of field $f(\mathbf{R}, t)$ can be obtained by additionally considering its spatial gradient $\mathbf{p}(\mathbf{R}, t) = \nabla f(\mathbf{R}, t)$. So, for example, the quantity

$$l(t; f) = \oint dl = \int d\mathbf{R}\, |\mathbf{p}(\mathbf{R}, t)|\, \delta(f(\mathbf{R}, t) - f) \qquad (8.7)$$

describes the total length of contours. The integrand in Eq. (8.7) is described by the extended indicator function

$$\varphi(\mathbf{R}, t; f, \mathbf{p}) = \delta(f(\mathbf{R}, t) - f)\, \delta(\mathbf{p}(\mathbf{R}, t) - \mathbf{p}), \qquad (8.8)$$

and the mean value of $l(t; f)$ (see formula (8.7)) is related to the joint one-time probability density of field $f(\mathbf{R}, t)$ and its gradient $\mathbf{p}(\mathbf{R}, t)$, which is obtained by averaging the indicator function (8.8) over an ensemble of realizations, which gives the function

$$P(\mathbf{R}, t; f, \mathbf{p}) = \langle \delta(f(\mathbf{R}, t) - f)\, \delta(\mathbf{p}(\mathbf{R}, t) - \mathbf{p}) \rangle.$$

By additionally considering spatial derivatives of the second order, the total number of contours $f(\mathbf{R}, t) = f = \text{const}$ can be estimated with the help of approximate formula (up to not-closing lines)

$$N(t; f) = N_{in}(t; f) - N_{out}(t; f)$$

$$= \frac{1}{2\pi} \int d\mathbf{R} \, \kappa(t, \mathbf{R}; f) \, |\mathbf{p}(\mathbf{R}, t)| \, \delta \, (f(\mathbf{R}, t) - f), \quad (8.9)$$

where $N_{in}(t; f)$ $(N_{out}(t; f))$ is the number of contours for which the vector $\mathbf{p}$ is directed along the inner (outer) normal, and $\kappa(t, \mathbf{R}; f)$ is the curvature of the level line.

8.2.1 Conditions of Cluster Structure Formation

We now discuss the conditions of stochastic structure formation in parametrically excited random fields. It is clear that for a *positive field* $f(\mathbf{R}, t)$ *in general* the condition of cluster formation with a unit probability, i.e., for almost all realizations, is the simultaneous tendency of being fulfilled, as $t \to \infty$, for asymptotic equalities

$$\langle S(t; f) \rangle \to 0, \quad \langle M(t; f) \rangle \to \int d\mathbf{R} \, \langle f(\mathbf{R}, t) \rangle \, .$$

The lack of structure formation corresponds to the simultaneous tendency towards satisfying the asymptotic equalities as $t \to \infty$:

$$\langle S(t; f) \rangle \to \infty, \quad \langle M(t; f) \rangle \to \int d\mathbf{R} \, \langle f(\mathbf{R}, t) \rangle \, .$$

For a spatially homogeneous field $f(\mathbf{R}, t)$, the one-point probability density $P(\mathbf{R}, t; f)$ does not depend on $\mathbf{R}$; in this case, statistical means of all the expressions (without integration over $\mathbf{R}$) will describe specific (per unit area) values of the respective quantities.

Thus, the specific mean area $\langle s_{hom}(t; f) \rangle$, where the random field $f(\mathbf{R}, t)$ exceeds a given level f, coincides with the event probability at any point in space, $f(\mathbf{R}, t) > f$:

$$\langle s_{hom}(t; f) \rangle = \langle \theta(f(\mathbf{R}, t) - f) \rangle = P\{f(\mathbf{R}, t) > f\}$$

so that the mean specific area offers a geometric interpretation of the probability of the event $f(\mathbf{R}, t) > f$, which is, clearly, independent of point $\mathbf{R}$. As a consequence, the conditions of clustering for the *homogeneous* case reduce to the tendency of being valid, as $t \to \infty$:

$$\langle s_{hom}(t; f) \rangle = P\{f(\mathbf{r}, t) > f\} \to 0, \quad \langle m_{hom}(t; f) \rangle \to \langle f(t) \rangle \, ,$$

whereas the absence of clustering is linked to the tendency of being valid, as $t \to \infty$:

$$\langle s_{hom}(t; f) \rangle = P\{f(\mathbf{r}, t) > f\} \to 1, \quad \langle m_{hom}(t; f) \rangle \to \langle f(t) \rangle \, .$$

Thus, *clustering in a spatially homogeneous problem is a phenomenon (occurring with a unit probability, i.e., for almost all realizations of a random positive field) spawned by a rare event occurring with vanishing probability.*

In this case, the mere presence of rare events serves as a *trigger* which initiates the structure formation process, while the structure formation on its own is the intrinsic property of a random medium, i.e., is in essence the *law of Nature* [30–34].

The characteristic time of cluster structure formation in space is determined by the character of the asymptotic expressions given above at large times. Now, this time is defined not only by the Lyapunov statistical characteristic index α, but also by the diffusion coefficient D_f in phase space of a positive field $f(\mathbf{r}, t)$. It is certainly larger than the characteristic time of realization decay at any fixed point in space.

For concrete physical dynamic systems, the description of clustering in physical fields reduces, therefore, to computation of the stochastic Lyapunov index α and diffusion coefficient D_f, which is, generally speaking, a rather tedious task for concrete partial differential equations.

In the presence of clustering, the field is simply absent over a large portion of space! It should be clear that the conditions above on the presence or absence of clustering in the field $f(\mathbf{R}, t)$ have nothing to do with the parametric growth of statistical characteristics like moment or correlation functions of arbitrary order, as time progresses.

The above criterion of ideal clustering (in analogy with ideal fluid dynamics) corresponds to the dynamics of cluster formation in dynamic systems governed, generally speaking, by partial differential equations of the first order (the Eulerian representation). This ideal structure emerges as a very narrow band (in the two-dimensional case) or very narrow tubes (in the three-dimensional case).

Notice that the first-order partial differential equations can be solved in general by the methods of characteristics. This corresponds to the Lagrangian description of dynamic systems. In this case, the characteristic curves described by ordinary differential equations can, naturally, have different peculiarities and even singularities. The conditions for the appearance of such peculiarities in the Lagrangian description do not have direct connections to the phenomenon of clustering in space and time, i.e., in the Eulerian representation.

However, in real physical systems, various additional factors related to the generation of spatial derivatives of a random field may become visible later; they *distort* but *do not eliminate* this picture of clustering. In particular, a situation is possible where the respective probability density approaches steady-state regime $P(\mathbf{R}; f)$ as $t \to \infty$. In this case, the functional of the form

$$\langle S(f) \rangle = \int d\mathbf{R} \int_f^\infty df' \, P(\mathbf{R}; f') \quad \text{and} \quad \langle M(f) \rangle = \int d\mathbf{R} \int_f^\infty df' \, f' P(\mathbf{R}; f')$$

cease to describe further distortions of the cluster picture. We need to study the temporal evolution of the functionals related to the spatial derivatives of the field $f(\mathbf{R}, t)$, such as the total contour length and the number of contours.

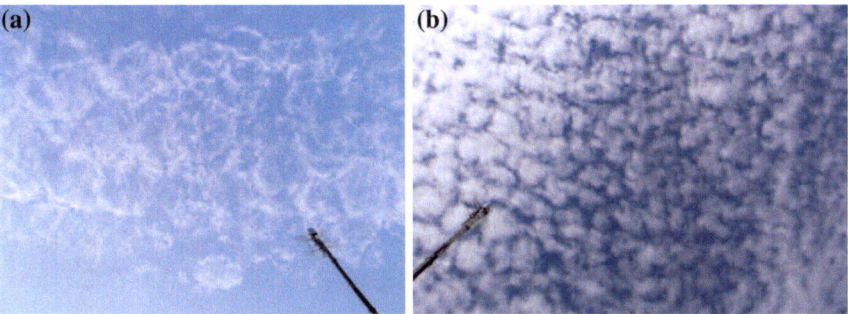

Fig. 8.2 Photos of the cluster structure in a cloudy sky: **a** cluster structure close to an 'ideal' one, and **b** 'deformed' clusters. The black stripes in the bottom corners are parts of a ground-based antenna. Photos are taken on 15 June and 2 August 2013 at 21:00 on the coast of the Sea of Azov. (Courtesy of V.A. Dovzhenko.)

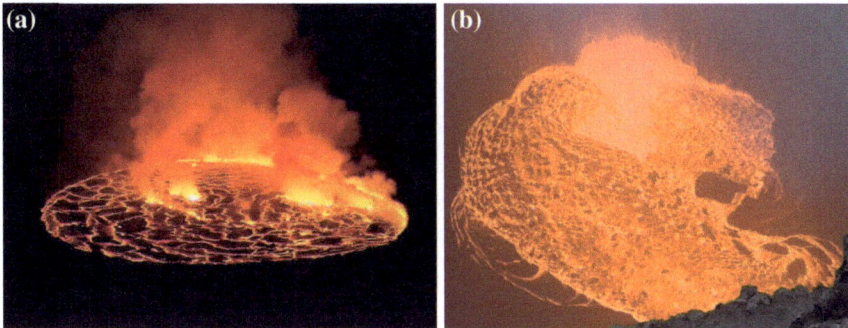

Fig. 8.3 **a** A lake of boiling lava in the Nyiragongo Crater in the Great Lakes region of Africa. **b** Lava lake in Halema'uma'u Crater on Kilauea (Photo courtesy of Hawaiian Volcano Observatory, USGS). (These images can be found on the sites http://bigpicture.ru/?p=128340, http://pacificislandparks. com/2010/01/20/more-amazing-lava-lake-photos/)

As an example of 'ideal' and 'deformed' clustering in Nature, we present here photos of the cluster structure of a cloudy sky[1] (Fig. 8.2) and a lava lake in volcano craters (Fig. 8.3).

We note that a statistical theory describing volcanic lava dynamics, as far as we know, does not exist at all.

A similar pattern is demonstrated by Figs. 8.4 and 8.5 related to the parabolic Leontovich (Schrödinger) equation (3.47) (see, for example, monographs [28, 29]).

To begin with, we note that if a plane wave propagates in a random medium, then in the delta-correlated approximation of random field $\varepsilon(x, \mathbf{R})$, in the longitudinal direction, and in which its correlation function $B_\varepsilon(x - x', \mathbf{R} - \mathbf{R}') = \langle \varepsilon(x, \mathbf{R})\varepsilon(x', \mathbf{R}') \rangle$ has the form

---

[1] Yet there is no particle clustering in the Lagrangian description!

(a)   **(b)**

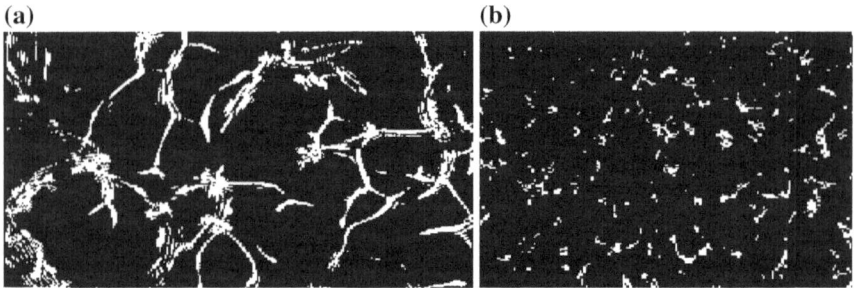

Fig. 8.4 Results of numerical modeling with the help of a system of phase screens (3.50): transverse section of a laser beam propagating through a turbulent medium in the region of strong focusing (**a**), and in the region of strong (saturated) fluctuations (**b**)

$$B_\varepsilon(x, \mathbf{R}) = A(\mathbf{R})\delta(x), \quad A(\mathbf{R}) = \int\limits_{-\infty}^{\infty} dx \, B_\varepsilon(x, \mathbf{R}),$$

Equation (3.47) lead for the mean field $\langle u(x, \mathbf{R}) \rangle$ and the second-order coherence function

$$\Gamma_2(x, \rho) = \left\langle u\left(x, \mathbf{R} + \frac{1}{2}\rho\right) u^*\left(x, \mathbf{R} - \frac{1}{2}\rho\right) \right\rangle,$$

to the expressions

$$\langle u(x, \mathbf{R}) \rangle = u_0 e^{-\gamma x/2}, \quad \Gamma_2(x, \rho) = |u_0|^2 e^{-k^2 x D(\rho)/4},$$

which do not depend on the wave field diffraction; here, $\gamma = (k^2/4)A(0)$ is the *extinction coefficient*, and the function $D(\rho) = A(0) - A(\rho)$ is linked to the structure function of the random field $\varepsilon(x, \mathbf{R})$. Simultaneously, a statistical scale ρ_{cog} called the *coherence radius* of field $u(x, \mathbf{R})$ appears in the plane perpendicular to the wave propagation direction, which is defined by the condition $(1/4)k^2 x D\left(\rho_{\text{cog}}\right) = 1$. The coherence radius depends on the wavelength, the distance travelled by the wave, and the medium statistical parameters.

Notice that a complete solution to the problem of computation of the statistical characteristics of solutions to Eq. (3.47) for $x \to \infty$ was found already in 1977 in Ref. [59] (see also Refs. [28, 29, 60]) by resorting to the continual integral representation of solutions to Eq. (3.47).

Obviously, the probability distribution of the wave field intensity has a lognormal character if the distance travelled by the incident plane wave remains small, and in this case the stochastic structure formation (clustering) ensues.

As the distance increases, the nonlinear character of the equation for the complex-valued phase needs to be taken into account. This region of fluctuations, referred to as *the region of strong focusing*, is extremely difficult for analytical analyses. For even larger distances of wave propagation, the statistical characteristics of wave intensity

(a)   **(b)**

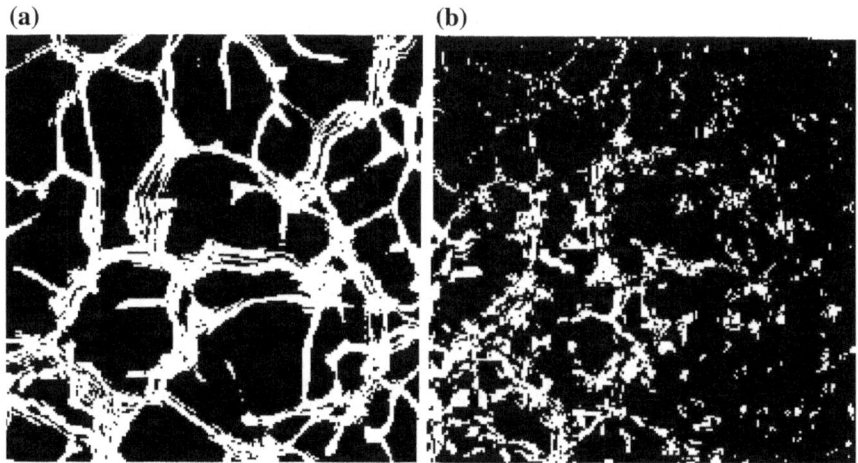

Fig. 8.5 Transverse section of a laser beam passing through a turbulent medium (under laboratory conditions) in the region of strong focusing (**a**), and in the region of strong (saturated) fluctuations (**b**)

reach a saturated regime; the respective spatial domain is referred to as the *region of strong intensity fluctuations*.

In this region, the statistical characteristics of wave field intensity cease to depend on the distance and take the form ($u_0 = 1$)

$$\langle I^n(x, \mathbf{R}) \rangle = n!, \quad P(x, I) = e^{-I}.$$

Reference [59] computed the spatial correlation function of wave field intensity $I(x, \mathbf{R}) = |u(x, \mathbf{R})|^2$ for $x \to \infty$ ($\rho = \mathbf{R}' - \mathbf{R}''$):

$$B_I(x, \rho) = \langle I(x, \mathbf{R}')I(x, \mathbf{R}'') \rangle - 1 = |\Gamma_2(x, \rho)|^2 = \exp\left\{-\frac{k^2 x}{2} D(\rho)\right\}, \quad (8.10)$$

which is also independent of the wave field diffraction. Now, in this problem, in addition to the spatial scale ρ_{cog}, a second characteristic spatial scale appears $r_0 = \dfrac{x}{k\rho_{\text{cog}}}$. However, numerous attempts by experimentalists, continuing into the present, to associate these scales with the patterns displayed in Figs. 8.4 and 8.5 have not led to success. And it is clear why! Clustering of the wave field intensity is certainly affected by diffraction, which is, however, in no way reflected in the form of its correlation function (8.10).

From the standpoint of statistical topography, the mean specific area of regions inside which $I(x, \mathbf{R}) > I$, and the mean specific power confined there are constant and in the limit of $x \to \infty$ do not describe the behavior of wave field intensity in individual realizations. Besides, no information is gained in this case by transition to the statistically equivalent random process. For this case, an explanation of the wave field structure in individual realizations was proposed only 20 years later (in 1997) in Ref. [61] (see also Refs. [28, 29]), by resorting to the analysis of such quantities as the specific mean length of contours and specific mean number of contours of wave field intensity, which are described by functionals like (8.7) and (8.9) and dependent on spatial derivatives of wave field intensity. These functions continue to increase with distance also in the region of strong intensity fluctuations, and, consequently, contour fragmentation takes place, as observed in laboratory experiments and through numerical modeling as well.

8.2.2   Statistical Topography of Lognormal Random Fields

In the analysis of one-point statistical characteristics in the spatially homogeneous case, it is generally rewarding to take into account that the random field $f(\mathbf{R}, t)$ is statistically equivalent to some random process $f(t)$ with the same statistical characteristics.

If a one-point probability density of the random field $f(\mathbf{r}, t)$ (8.2) is known, one can also obtain general information on the spatial structure of random field $f(\mathbf{r}, t)$. In particular, such functional of the random field $f(\mathbf{r}, t)$, as the common mean volume (in three dimensions) or area (in two dimensions) of the region, where $f(\mathbf{r}, t) > f$, and the common mean 'mass' of the field comprised there, are described as

$$\langle V(t, f) \rangle = \int d\mathbf{r} \int_f^\infty df' \, P(\mathbf{r}, t; f'), \quad \langle M(t, f) \rangle = \int d\mathbf{r} \int_f^\infty df' \, f' P(\mathbf{r}, t; f').$$

The values of these functionals do not depend on diffusion in the $\mathbf{r}$-space (the coefficient D_0), and for probability distribution (8.2) we find the expressions

$$\langle V(t, f) \rangle = \int d\mathbf{r} \, \mathrm{Pr} \left(\frac{1}{\sqrt{2Dt}} \ln \left(\frac{f_0(\mathbf{r})}{f} e^{-\alpha t} \right) \right),$$

$$\langle M(t, f) \rangle = e^{(D-\alpha)t} \int d\mathbf{r} \, f_0(\mathbf{r}) \, \mathrm{Pr} \left(\frac{1}{\sqrt{2Dt}} \ln \left(\frac{f_0(\mathbf{r})}{f} e^{(2D-\alpha)t} \right) \right), \qquad (8.11)$$

where the probability integral $\mathrm{Pr}(z)$ is defined by equality (5.4).

Taking now into account the asymptotics of function $\mathrm{Pr}(z)$ (5.5), one can analyze how functionals (8.11) evolve with time. Namely, for $t \to \infty$, the asymptotics of the mean volume decays with time according to the law

$$\langle V(t, f) \rangle \approx \frac{1}{\alpha} \sqrt{\frac{D}{\pi f^{\alpha/D} t}} e^{-\alpha^2 t/4D} \int d\mathbf{r} \sqrt{f_0^{\alpha/D}(\mathbf{r})}$$

for $\alpha > 0$. For $\alpha < 0$, the mean volume occupies the entire space as $t \to \infty$.

For the common mean 'mass', asymptotics in the limit $t \to \infty$ has the form (in the most interesting case $\alpha < 2D$)

$$\langle M(t, f) \rangle \approx e^{(D-\alpha)t} \int d\mathbf{r}\, f_0(\mathbf{r}) \times$$

$$\times \left[1 - \frac{1}{(2D-\alpha)} \sqrt{\frac{D}{\pi t}} \left(\frac{f}{f_0(\mathbf{r})} \right)^{(2D-\alpha)/D} e^{-(2D-\alpha)^2 t/4} \right].$$

As a consequence, for $\alpha > 0$, clusters contain the overall mean 'mass' in the limit $t \to \infty$.

For homogeneous initial conditions, the respective expressions taken without integration over $\mathbf{r}$ describe the specific values of the volume comprising large excursions and their common 'mass' per unit volume, i.e.

$$\langle v_{\text{hom}}(t, f) \rangle = \langle \theta \left(f(\mathbf{r}, t) - f \right) \rangle = \mathsf{P}\{ f(\mathbf{r}, t) > f \} = \mathrm{Pr} \left(\frac{1}{\sqrt{2Dt}} \ln \left(\frac{f_0}{f} e^{-\alpha t} \right) \right),$$

$$\langle m_{\text{hom}}(t, f) \rangle = f_0 e^{(D-\alpha)t} \, \mathrm{Pr} \left(\frac{1}{\sqrt{2Dt}} \ln \left(\frac{f_0}{f} e^{(2D-\alpha)t} \right) \right). \tag{8.12}$$

If we select a section level $f > f_0$, then at the initial instant of time $\langle v_{\text{hom}}(0, f) \rangle = 0$ and $\langle m_{\text{hom}}(0, f) \rangle = 0$. Spatial perturbations of the random field $f(\mathbf{r}, t)$ evolve later, and in the limit $t \to \infty$ we arrive at the asymptotic expressions $(2D > \alpha)$

$$\langle v_{\text{hom}}(t, f) \rangle = \mathsf{P}\{ f(\mathbf{r}, t) > f \} \approx$$

$$\approx \begin{cases} \dfrac{1}{\alpha} \sqrt{\dfrac{D}{\pi t}} \left(\dfrac{f_0}{f} \right)^{\alpha/D} e^{-\alpha^2 t/4D} \quad (\alpha > 0), \\[4mm] 1 - \dfrac{1}{|\alpha|} \sqrt{\dfrac{D}{\pi t}} \left(\dfrac{f}{f_0} \right)^{|\alpha|/D} e^{-\alpha^2 t/4D} \quad (\alpha < 0), \end{cases} \tag{8.13}$$

$$\langle m_{\text{hom}}(t, f) \rangle \approx f_0 e^{(D-\alpha)t} \left[1 - \frac{1}{(2D-\alpha)} \sqrt{\frac{D}{\pi t}} \left(\frac{f}{f_0} \right)^{(2D-\alpha)/D} e^{-(2D-\alpha)^2 t/4D} \right].$$

$$\tag{8.14}$$

Thus, for $\alpha > 0$ the specific common volume tends to zero, while the specific common 'mass' confined within it tends to the mean 'mass' in the entire space. This corresponds to the criterion of structure formation with a unit probability through 'ideal clustering' of the field $f(\mathbf{r}, t)$ being considered. In this case, *the random field $f(\mathbf{r}, t)$*

is practically absent in the dominant part of the space. The field characteristic decay
time at any fixed point in space can be estimated as $\alpha t \sim 1$, and the characteristic
time of field cluster structure formation as $\alpha t \sim \max\left\{4\xi,\ 4\xi/(2\xi - 1)^2\right\}$, where
$\xi = D/\alpha$.

If $\alpha < 0$, clustering is lacking and only general amplification of the random field
$f(\mathbf{r}, t)$ takes place everywhere in space. Thus, *chaos remains chaos in that case!*
Only zeros of field $f(\mathbf{r}, t)$ cluster.

Let us mention that the following theorem holds.

*A conservative positive parametrically excited lognormal field in a statistically
homogeneous case always produces clusters with a unit probability, i.e., for almost
all its realizations.*

Indeed, in this case, $f(\mathbf{r}, t) = e^{\ln f(\mathbf{r}, t)}$; hence, one has

$$\langle f(\mathbf{r}, t) \rangle = \left\langle e^{\ln f(\mathbf{r}, t)} \right\rangle = \exp\left\{ \langle \ln f(\mathbf{r}, t) \rangle + \frac{1}{2}\sigma^2_{\ln f(\mathbf{r}, t)} \right\},$$

where $\sigma^2_{\ln f(\mathbf{r}, t)}$ is the variance of the random field $\ln f(\mathbf{r}, t)$. Now taking into account
that conservative character of this field leads to

$$\langle \ln f(\mathbf{r}, t) \rangle + \frac{1}{2}\sigma^2_{\ln f(\mathbf{r}, t)} = \ln f_0,$$

we find for the typical realization curve

$$f^*(\mathbf{r}, t) = e^{\langle \ln f(\mathbf{r}, t) \rangle} = f_0 \exp^{-\alpha t},$$

where the Lyapunov characteristic parameter

$$\alpha = \lim_{t \to \infty} \frac{1}{2t}\sigma^2_{\ln f(\mathbf{r}, t)} > 0$$

and the problem is that of computing it from the respective dynamic equation. Since,
as pointed out earlier, for the conservative field $f(\mathbf{r}, t)$ the parameter $\alpha = D$ (see
Eq. (8.3)), the characteristic time of cluster structure formation is $\alpha t \sim 4$, which is
four times larger than the characteristic field decay time at almost any point in space.

For instance, for cluster formation in a density field, we have dynamic equation
(3.51).

With complex-valued parabolic equation (3.47), we have the equation for the wave
field intensity — the continuity equation (3.48) and, consequently, introducing the
amplitude level as $\chi(\mathbf{R}, x) = \ln A(\mathbf{R}, x)$, we find for a plane wave the Lyapunov
exponent in the form

$$I^*(\mathbf{R}, x) = I_0 e^{-2\alpha x},$$

where the parameter $\alpha = \lim\limits_{x\to\infty} \dfrac{1}{x}\sigma^2_{\chi(\mathbf{R},x)}$, and $\sigma^2_{\chi(\mathbf{R},x)}$ is the variance of the amplitude level computed in the framework of the first approximation of the smooth perturbation method proposed by S.M. Rytov (see, for example, books [28, 29]).

As an application of the theory elaborated here, we consider concrete physical stochastic transport phenomena in random media spawned through the parametric action of the medium on the respective dynamic system.

Chapter 9
Stochastic Transport Phenomena in a Random Velocity Field

...Chaos is the place which serves to contain all things;
for if this had not subsisted neither earth nor water
nor the rest of the elements, nor the Universe
as a whole, could have been constructed. ...

Sextus Empiricus, *Against the Physics, against the Ethicists,*
R.G. Bury, p. 217, Harvard University Press, 1997.

9.1 Clustering of the Density Field in a Random Velocity Field

Stochastic structure formation in a spatially homogeneous statistical problem on diffusion of the density field $\rho(\mathbf{r}, t)$ in a random velocity field is described by Eq. (3.51) that we rewrite in the form

$$\left(\frac{\partial}{\partial t} + \mathbf{u}(\mathbf{r}, t)\frac{\partial}{\partial \mathbf{r}}\right)\rho(\mathbf{r}, t) + \frac{\partial \mathbf{u}(\mathbf{r}, t)}{\partial \mathbf{r}}\rho(\mathbf{r}, t) = 0, \quad \rho(\mathbf{r}, 0) = \rho_0(\mathbf{r}). \quad (9.1)$$

To describe the density field in the Eulerian description, we introduce the indicator function:

$$\varphi(\mathbf{r}, t; \rho) = \delta(\rho(\mathbf{r}, t) - \rho), \quad (9.2)$$

which is localized on surface $\rho(\mathbf{r}, t) = \rho = \text{const}$ in the three-dimensional case or on a contour in the two-dimensional case. Differentiating Eq. (9.2) with respect to time t and using dynamic equation (9.1) and 'probing property' of the delta-function, we obtain the equation

$$\frac{\partial}{\partial t}\varphi(\mathbf{r}, t; \rho) = \frac{\partial \mathbf{u}(\mathbf{r}, t)}{\partial \mathbf{r}}\frac{\partial}{\partial \rho}\rho\varphi(\mathbf{r}, t; \rho) + \mathbf{u}(\mathbf{r}, t)\frac{\partial \rho(\mathbf{r}, t)}{\partial \mathbf{r}}\frac{\partial}{\partial \rho}\varphi(\mathbf{r}, t; \rho). \quad (9.3)$$

© Springer International Publishing AG 2017
V.I. Klyatskin, *Fundamentals of Stochastic Nature Sciences,*
Understanding Complex Systems, DOI 10.1007/978-3-319-56922-2_9

However, this equation is unclosed because the right-hand side includes term $\partial \rho(\mathbf{r}, t)/\partial \mathbf{r}$ that cannot be explicitly expressed through $\rho(\mathbf{r}, t)$.

On the other hand, differentiating function (9.2) with respect to $\mathbf{r}$, we obtain the equality

$$\frac{\partial}{\partial \mathbf{r}} \varphi(\mathbf{r}, t; \rho) = -\frac{\partial \rho(\mathbf{r}, t)}{\partial \mathbf{r}} \frac{\partial}{\partial \rho} \varphi(\mathbf{r}, t; \rho). \tag{9.4}$$

Eliminating now the last term in Eq. (9.3) with the use of (9.4), we obtain the closed Liouville equation in the Eulerian description:

$$\left(\frac{\partial}{\partial t} + \mathbf{u}(\mathbf{r}, t) \frac{\partial}{\partial \mathbf{r}} \right) \varphi(\mathbf{r}, t; \rho) = \frac{\partial \mathbf{u}(\mathbf{r}, t)}{\partial \mathbf{r}} \frac{\partial}{\partial \rho} [\rho \varphi(\mathbf{r}, t; \rho)], \tag{9.5}$$
$$\varphi(\mathbf{r}, 0; \rho) = \delta(\rho_0(\mathbf{r}) - \rho).$$

Averaging Eq. (9.4) over an ensemble of realizations of field $\mathbf{u}(\mathbf{r}, t)$, we obtain for the field $\rho(\mathbf{r}, t)$ probability density $P(\mathbf{r}, t; \rho) = \langle \delta (\rho(\mathbf{r}, t) - \rho) \rangle_\mathbf{u}$ the equation (see, for example, [28, 29])

$$\frac{\partial}{\partial t} P(\mathbf{r}, t; \rho) = D_0 \Delta P(\mathbf{r}, t; \rho) + D_\rho \frac{\partial^2}{\partial \rho^2} \rho^2 P(\mathbf{r}, t; \rho)$$

with initial condition $P(\mathbf{r}, 0; \rho) = \delta(\rho_0(\mathbf{r}) - \rho)$, where the diffusion coefficients D_0 (in $\mathbf{r}$-space) and $D_\rho = D^P$ (in ρ-space) are given by Eqs. (4.5) and (4.7).

In the case of uniform initial density field (if $\rho_0(\mathbf{r}) = \rho_0$) the probability distribution is independent of $\mathbf{r}$ and satisfies the equation.

$$\frac{\partial}{\partial t} P(t; \rho) = D_\rho \frac{\partial^2}{\partial \rho^2} \rho^2 P(t; \rho), \quad P(0; \rho) = \delta(\rho - \rho_0). \tag{9.6}$$

Equation (9.6) coincides with Eq. (8.4) at the fulfillment of equalities $\alpha = D = D_\rho = D^P$, where the quantity D^P is determined by the potential spectral component of the velocity field. Consequently, the one-point probability density of the density field is lognormal together with respective integral distribution function by the formulas

$$P(t; \rho) = \frac{1}{2\rho\sqrt{\pi\tau}} \exp \left\{ -\frac{\ln^2 (\rho e^\tau / \rho_0)}{4\tau} \right\},$$
$$\Phi(t; \rho) = \Pr \left(\frac{\ln (\rho e^\tau / \rho_0)}{2\sqrt{\tau}} \right), \tag{9.7}$$

where the parameter $\tau = D_\rho t$.

For one-point characteristics of the density field $\rho(\mathbf{r}, t)$, as already pointed out, the problem is statistically equivalent to the analysis of a random process, and in this case all the moment functions at any fixed point in space grow exponentially with time for both $n > 0$ and $n < 0$

$$\langle \rho(\mathbf{r}, t) \rangle = \rho_0, \quad \langle \rho^n(\mathbf{r}, t) \rangle = \rho_0^n e^{n(n-1)\tau}, \tag{9.8}$$

and the typical realization curve for the density field, which coincides with the Lyapunov exponent, exponentially decays with time at any fixed point in space:

$$\rho^*(t) = e^{\langle \ln(\rho(\mathbf{r}, t)) \rangle} = \rho_0 e^{-\tau},$$

which indicates that the density field decays with a unit probability (i.e., in almost all realizations of the density field) in arbitrary divergent flow at any fixed point in space. Then, the characteristic decay time for the density field $\tau \sim 1$. Note that Eq. (9.6) for the probability density corresponds to the Eulerian description of a density field. It should be recalled that in the Lagrangian description of the density field the system of characteristic curves (particles) (3.1) does not necessarily undergo clustering (see Sect. 5.2).

The formation of density field statistics at any fixed point in space (moment and correlation functions) occurs through density field fluctuations around the typical realization curve. Thus, in the case of a compressible flow (in a divergent velocity field) the density field always undergoes clustering with a unit probability. The specific mean area (volume) of the region where $\rho(\mathbf{r}, t) > \rho$, is expressed as

$$\langle s_{\text{hom}}(t, \rho) \rangle = \int_\rho^\infty d\rho' \, P(t; \rho') = P\{\rho(\mathbf{r}, t) > \rho\} = \Pr\left(\frac{\ln\left(\rho_0 \, e^{-\tau}/\rho\right)}{\sqrt{2\tau}}\right), \tag{9.9}$$

and the specific mean 'mass' of the tracer confined in this region is described as

$$\langle m_{\text{hom}}(t, \rho) \rangle_{\text{hom}} / \rho_0 = \frac{1}{\rho_0} \int_\rho^\infty \rho' \, d\rho' \, P(t; \rho) = \Pr\left(\frac{\ln\left(\rho_0 \, e^{\tau}/\rho\right)}{\sqrt{2\tau}}\right). \tag{9.10}$$

From Eqs. (9.9) and (9.10) it follows that for $\tau \gg 1$ the mean specific area (volume) decays according to the law

$$\langle s_{\text{hom}}(t, \rho) \rangle = P\{\rho(\mathbf{r}, t) > \rho\} \approx \sqrt{\frac{\rho_0}{\pi \rho \tau}} e^{-\tau/4}, \tag{9.11}$$

whereas almost all tracer 'mass' is accumulated within it:

$$\langle m_{\text{hom}}(t, \rho) \rangle / \rho_0 \approx 1 - \sqrt{\frac{\rho}{\pi \rho_0 \tau}} e^{-\tau/4}, \tag{9.12}$$

which corresponds to the physical phenomenon of density field clustering in a random velocity field. As follows from formulas (9.11) and (9.12), the characteristic time of cluster structure formation in the tracer field is four times that of the characteristic decay time for the density field at any fixed point in space ($\tau \sim 4$).

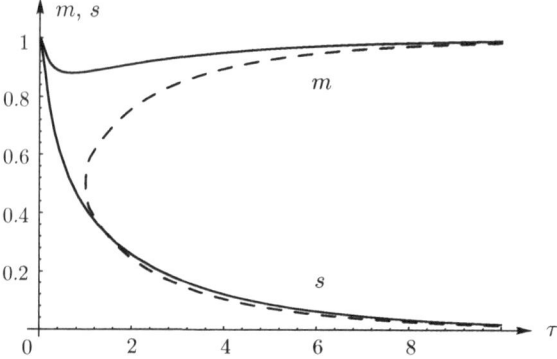

Fig. 9.1 Cluster formation dynamics for $\rho/\rho_0 = 0.5$

Nevertheless, the time-dependent behavior of the formation of cluster structure essentially depends on ratio ρ/ρ_0. If $\rho/\rho_0 < 1$, then $s(0, \rho) = 1$ and $m(0, \rho) = 1$ at the initial instant. Then, in view of the fact that particles of buoyant tracer initially tend to scatter, there appear small areas within which $\rho(\mathbf{r}, t) < \rho$ and which concentrate only insignificant portion of the total mass. These regions rapidly grow with time and their mass flows into cluster region relatively quickly approaching asymptotic expressions (9.11) and (9.12) (Fig. 9.1).

Note that $s(t^*, \rho) = 1/2$ at instant $t^* = \ln(\rho/\rho_0)$.

In the opposite, more interesting case $\rho/\rho_0 > 1$, we have $s(0, \rho) = 0$ and $m(0, \rho) = 0$ at the initial instant. In view of initial scatter of particles, there appear small cluster regions within which $\rho(\mathbf{r}, t) > \rho$; these regions remain at first almost invariable in time and intensively absorb a significant portion of total mass. With time, the area of these regions begins to decrease and the mass within them begins to increase according to asymptotic expressions (9.11) and (9.12) (Fig. 9.2a, b).

Note that even in an incompressible fluid, the density field will experience clustering in hydrodynamic flows if the tracer is 'buoyant', if the finite inertia of the density field is taken into account, and for multiphase fluid flows, i.e., always when a potential spectral component arises in the tracer velocity field, which is different from the velocity field of the fluid proper. This case corresponds, for example, to a cloudy sky (see Fig. 8.4). Here the nature of stochastic character of air masses is completely irrelevant—be it developed convection or atmospheric turbulence. Judging by the time the photo was taken we have the second case. But there is no particle clustering in the Lagrangian description!

Thus, for the continuity equation

$$\left(\frac{\partial}{\partial t} + \frac{\partial}{\partial \mathbf{r}} \mathbf{V}(\mathbf{r}, t)\right) \rho(\mathbf{r}, t) = 0, \quad \rho(\mathbf{r}, 0) = \rho_0(\mathbf{r}), \tag{9.13}$$

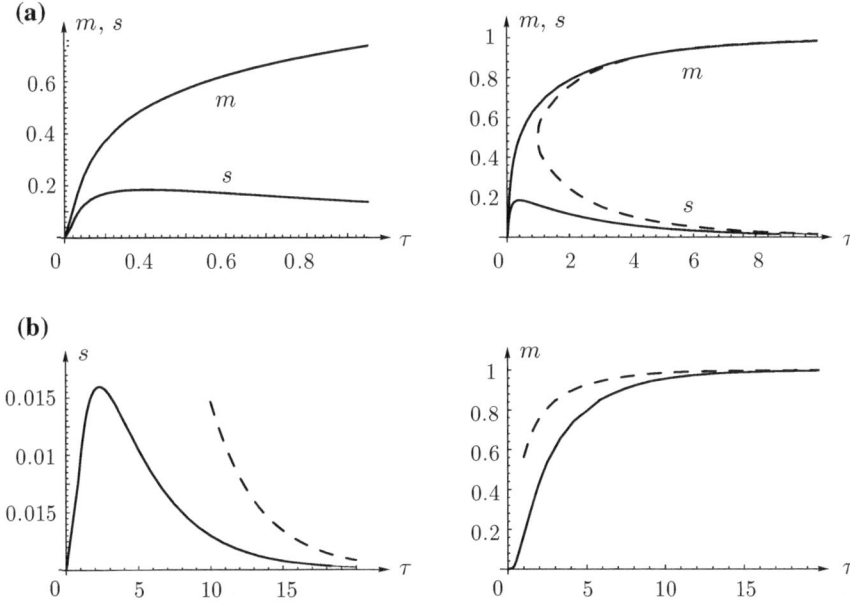

Fig. 9.2 Cluster formation dynamics for **a** $\rho/\rho_0 = 1.5$ and **b** $\rho/\rho_0 = 10$

describing the density field of a passive scalar tracer $\rho(\mathbf{r}, t)$, moving in a random hydrodynamic flow with the velocity $\mathbf{V}(\mathbf{r}, t)$ and having a potential component, tracer clustering always occurs with a unit probability independently of the dynamic equation governing the velocity field $\mathbf{V}(\mathbf{r}, t)$.

For example, in the case of tracer $\rho(\mathbf{r}, t)$ with low inertia, the velocity field $\mathbf{V}(\mathbf{r}, t)$ can be described by the phenomenological equation (see Ref. [62])

$$\left(\frac{\partial}{\partial t} + \mathbf{V}(\mathbf{r}, t)\frac{\partial}{\partial \mathbf{r}}\right) \mathbf{V}(\mathbf{r}, t) = -\lambda\left[\mathbf{V}(\mathbf{r}, t) - \mathbf{u}(\mathbf{r}, t)\right], \qquad (9.14)$$

where $\mathbf{u}(\mathbf{r}, t)$ is the velocity field of the hydrodynamic flow itself, and the parameter $\tau = 1/\lambda$ is the known Stokes time depending on the size of tracer particles and molecular viscosity. Equation (9.14) is that of a simple wave with linear friction and a random force related to the hydrodynamic flow. A peculiarity of this equation consists in the fact that it is valid only in the asymptotic limit $\lambda \to \infty$. This implies that the parameter $\lambda\tau_0 \gg 1$, where τ_0 to is the temporal correlation radius for the hydrodynamic velocity $\mathbf{u}(\mathbf{r}, t)$, and, consequently, the approximation of field $\mathbf{u}(\mathbf{r}, t)$ delta-correlated in time is inapplicable for the statistical problem description, and the finiteness of the temporal correlation radius of the field τ_0 needs to be taken into account.

Under the assumption that the variance of the random velocity field $\sigma_{\mathbf{u}}^2 = \langle \mathbf{u}^2(\mathbf{r}, t)\rangle$ is sufficiently small for large values of parameter λ (low inertia of particles),

Eq. (9.14) can be linearized with respect to the function $\mathbf{V}(\mathbf{r}, t) \approx \mathbf{u}(\mathbf{r}, t)$ and we then arrive at a simpler vector equation

$$\left(\frac{\partial}{\partial t} + \mathbf{u}(\mathbf{r}, t)\frac{\partial}{\partial \mathbf{r}}\right) \mathbf{V}(\mathbf{r}, t) = -\left(\mathbf{V}(\mathbf{r}, t)\frac{\partial}{\partial \mathbf{r}}\right) \mathbf{u}(\mathbf{r}, t) - \lambda\left[\mathbf{V}(\mathbf{r}, t) - \mathbf{u}(\mathbf{r}, t)\right].$$

In this approximation, the probability density of the density field is described by an equation like Eq. (7.60)

$$\frac{\partial}{\partial t} P(\mathbf{r}, t; \rho) = \left(D_0 \frac{\partial^2}{\partial \mathbf{r}^2} + D_\rho \frac{\partial^2}{\partial \rho^2}\rho^2\right) P(\mathbf{r}, t; \rho),$$
$$P(\mathbf{r}, 0; \rho) = \delta\left(\rho_0(\mathbf{r}) - \rho\right),$$
(9.15)

where D_0 and D_ρ are the diffusion coefficients:

$$D_0 = \frac{1}{d}\int_0^\infty d\tau \, \langle \mathbf{V}(\mathbf{r}, t + \tau)\mathbf{V}(\mathbf{r}, t)\rangle = \frac{1}{d}\tau_\mathbf{V}\langle \mathbf{V}^2(\mathbf{r}, t)\rangle,$$
$$D_\rho = \int_0^\infty d\tau \left\langle \frac{\partial \mathbf{V}(\mathbf{r}, t + \tau)}{\partial \mathbf{r}} \frac{\partial \mathbf{V}(\mathbf{r}, t)}{\partial \mathbf{r}}\right\rangle = \tau_{\mathrm{div}\mathbf{V}}\left\langle\left(\frac{\partial \mathbf{V}(\mathbf{r}, t)}{\partial \mathbf{r}}\right)^2\right\rangle,$$
(9.16)

where $\tau_\mathbf{V}$ and $\tau_{\mathrm{div}\mathbf{V}}$ are the temporal correlation radii for the random fields $\mathbf{V}(\mathbf{r}, t)$ and $\partial \mathbf{V}(\mathbf{r}, t)/\partial \mathbf{r}$.

For an incompressible fluid flow in the diffusion approximation, diffusion coefficients (9.16) are described by the expressions [62]

$$D_0 = \frac{1}{d}\tau_\mathbf{V}\langle \mathbf{V}^2(\mathbf{r}, t)\rangle = \frac{1}{d}\tau_0 B_{ii}(0, 0) = \frac{d-1}{d}\tau_0 \int d\mathbf{k} E(k, 0),$$
$$D_\rho = \tau_{\mathrm{div}\mathbf{V}}\left\langle\left(\frac{\partial \mathbf{V}(\mathbf{r}, t)}{\partial \mathbf{r}}\right)^2\right\rangle = \frac{4}{\lambda}\frac{d^2-1}{d(d+2)}D_1 D_2(\lambda),$$
(9.17)

where the coefficient

$$D_1 = -\frac{\tau_0}{(d-1)}\langle \mathbf{u}(\mathbf{r}, t)\Delta\mathbf{u}(\mathbf{r}, t)\rangle$$

does not depend on the parameter λ. The coefficient $D_2(\lambda)$, if $\lambda\tau_0 \gg 1$, is defined by the expression

$$D_2(\lambda) = -\frac{1}{\lambda(d-1)}\langle \mathbf{u}(\mathbf{r}, t)\Delta\mathbf{u}(\mathbf{r}, t)\rangle.$$

Thus, we see that the coefficient D_ρ in Eq. (9.15) is proportional $\sigma_\mathbf{u}^4$. And the vortex component of field $\mathbf{u}(\mathbf{r}, t)$ first generates the vortex component of the field $\mathbf{V}(\mathbf{r}, t)$ by a direct linear mechanism without a contribution from advection, and only then do the vortex component of the field $\mathbf{V}(\mathbf{r}, t)$ generate a divergent component of the field $\mathbf{V}(\mathbf{r}, t)$ through the advection mechanism.

Note that for particles with low inertia in the presence of buoyancy and gravity forces the tracer velocity field $\mathbf{V}(\mathbf{r}, t)$ in the hydrodynamic flow $\mathbf{u}(\mathbf{r}, t)$, is described by the equations (see, for example, Ref. [63])

$$\left(\frac{\partial}{\partial t} + \mathbf{V}(\mathbf{r}, t)\frac{\partial}{\partial \mathbf{r}}\right)\mathbf{V}(\mathbf{r}, t) = -\lambda\left[\mathbf{V}(\mathbf{r}, t) - \mathbf{u}(\mathbf{r}, t)\right] + \mathbf{g}\left(1 - \frac{\rho_0}{\rho_P}\right), \quad (9.18)$$

where g is the gravitational acceleration, and ρ_P and ρ_0 are the densities of tracer particles and the medium, respectively.

The velocity $\mathbf{v}$, of tracer sedimentation or floating up, directed, as a rule, vertically, is determined by the balance between the buoyancy and viscous friction forces for a moving tracer and is described by the formula

$$\frac{\mathbf{g}}{\lambda}\left(1 - \frac{\rho_0}{\rho_P}\right) = \mathbf{v}.$$

Writing now $\mathbf{V}(\mathbf{r}, t) = \mathbf{v} + \mathbf{v}(\mathbf{r}, t)$, where $\mathbf{v}(\mathbf{r}, t)$ are the fluctuations of a tracer velocity field with respect to $\mathbf{v}$, for the system of equations (9.13) and (9.18) the one-point probability density $P(\mathbf{r}, t; \rho)$ is described by Eq. (9.15), where the diffusion coefficient $D_\rho(\mathbf{v})$, depending on the sedimentation velocity $\mathbf{v}$, is now given by the expression [64]

$$D_\rho(\mathbf{v}) = \frac{4(d+1)}{d(d+2)(d-1)\lambda^2} \frac{\partial^2 B_{\alpha\alpha}^{(\mathbf{u})}(\mathbf{0}, 0)}{\partial \mathbf{r}^2} \int\limits_0^\infty d\tau \frac{\partial^2 B_{\beta\beta}^{(\mathbf{u})}(\mathbf{v}\tau, \tau)}{\partial \mathbf{r}^2}. \quad (9.19)$$

Consequently, the presence of tracer sedimentation leads to a reduction in the diffusion coefficient $D_\rho(\mathbf{v})$, i.e., to a larger clustering time. In reality, the clustering of sedimenting weakly inertial particles explains numerous phenomena in Nature, for example, the spot structure of radioactive precipitation after the Chernobyl catastrophe (see, for example, Ref. [65]). The spotty structure of sand precipitation over the Indian Ocean after sand storms in African deserts is also well known.

We note that in review [66], dealing with the large-scale structure of the Universe, the vector equation of a simple wave (9.13), with no right-hand side but under the assumption that the velocity field $\mathbf{V}(\mathbf{r}, t)$ is potential, is 'glued' to the continuity equation (9.13). The randomness in this case is attributed to fluctuations in the initial conditions. Clearly, in this case the tracer field undergoes clustering with a unity probability in almost every its realization. However, the compliance of this fact with the observed distribution of galaxies can in no way be treated as confirmation of the relevance of the velocity field model used, based on the simple wave equation.

We stress that the clustering of a tracer field with a unit probability is realized for any model of a velocity field (linear or nonlinear) in the presence of a potential component in the velocity field.

In this section, we considered problems of stochastic structure formation of scalar density field in random velocity field within the first stage of the dynamics based on the continuity equation (3.51) neglecting the effect of the dynamic diffusion coefficient. The effect of this parameter on the stochastic structure formation appears for greater times and is described in terms of the second order partial differential equation (3.45). This case was considered in paper [67] by numerical modeling of the problem on buoyant tracer diffusion in random velocity field. The modeling showed that clustering of tracer occurs in this problem as before; however, here the contours appear to be fragmented, similar to those in the problem on propagation of laser radiation in a random medium considered earlier in Sect. 8.2.1.

9.2 Probabilistic Description of a Magnetic Field and Its Energy in a Random Velocity Field

9.2.1 Probabilistic Description of a Magnetic Field

Let us consider now a probabilistic description of a magnetic field based on dynamic equation (3.52) in a statistically homogeneous case. Just as for the density field, we will assume that the random component of a velocity field $\mathbf{u}(\mathbf{r}, t)$ is a divergent (div $\mathbf{u}(\mathbf{r}, t) \neq 0$) random Gaussian field, homogeneous and isotropic in space and stationary delta-correlated in time.

Introduce the indicator function of magnetic field $\mathbf{H}(\mathbf{r}, t)$

$$\varphi(\mathbf{r}, t; \mathbf{H}) = \delta(\mathbf{H}(\mathbf{r}, t) - \mathbf{H}).$$

Differentiating this function with respect to time with the use of dynamic equation (3.52) and probing property of the delta function, we arrive at the equation

$$\frac{\partial}{\partial t}\varphi(\mathbf{r}, t; \mathbf{H}) = -\frac{\partial}{\partial H_i}\left[\mathbf{H}\frac{\partial u_i(\mathbf{r}, t)}{\partial \mathbf{r}} - H_i\frac{\partial \mathbf{u}(\mathbf{r}, t)}{\partial \mathbf{r}}\right]\varphi(\mathbf{r}, t; \mathbf{H})$$
$$+ \mathbf{u}(\mathbf{r}, t)\frac{\partial H_i(\mathbf{r}, t)}{\partial \mathbf{r}}\frac{\partial}{\partial H_i}\varphi(\mathbf{r}, t; \mathbf{H}),$$

which is unclosed because of the last term. However, since

$$\frac{\partial}{\partial \mathbf{r}}\varphi(\mathbf{r}, t; \mathbf{H}) = -\frac{\partial H_i(\mathbf{r}, t)}{\partial \mathbf{r}}\frac{\partial}{\partial H_i}\delta(\mathbf{H}(\mathbf{r}, t) - \mathbf{H}),$$

and, hence,

$$-\mathbf{u}(\mathbf{r}, t)\frac{\partial}{\partial \mathbf{r}}\varphi(\mathbf{r}, t; \mathbf{H}) = \mathbf{u}(\mathbf{r}, t)\frac{\partial H_i(\mathbf{r}, t)}{\partial \mathbf{r}}\frac{\partial}{\partial H_i}\delta(\mathbf{H}(\mathbf{r}, t) - \mathbf{H}),$$

we obtain the desired closed Liouville equation [51]

$$\left(\frac{\partial}{\partial t} + \mathbf{u}(\mathbf{r}, t)\frac{\partial}{\partial \mathbf{r}}\right)\varphi(\mathbf{r}, t; \mathbf{H}) = -\frac{\partial}{\partial H_i}\left[\mathbf{H}\frac{\partial u_i(\mathbf{r}, t)}{\partial \mathbf{r}} - H_i\frac{\partial \mathbf{u}(\mathbf{r}, t)}{\partial \mathbf{r}}\right]\varphi(\mathbf{r}, t; \mathbf{H})$$

(9.20)

with the initial condition

$$\varphi(\mathbf{r}, 0; \mathbf{H}) = \delta(\mathbf{H}_0(\mathbf{r}) - \mathbf{H}).$$

The one-point probability density of magnetic field is defined as the equality

$$P(\mathbf{r}, t; \mathbf{H}) = \langle \varphi(\mathbf{r}, t; \mathbf{H})\rangle_{\mathbf{u}}.$$

Let us average Eq. (9.20) over an ensemble of realizations of field $\{\mathbf{u}(\mathbf{r}, t)\}$. As a result we obtain then the desired equation

$$\left(\frac{\partial}{\partial t} - D_0\frac{\partial^2}{\partial \mathbf{r}^2}\right)P(\mathbf{r}, t; \mathbf{H}) = \left\{D_1\frac{\partial}{\partial H_l}\frac{\partial}{\partial H_k}H_l H_k + D_2\frac{\partial}{\partial H_l}\frac{\partial}{\partial H_l}H_k^2\right\}P(\mathbf{r}, t; \mathbf{H}),$$

(9.21)

where D_1 and D_2 are the diffusion coefficients in the $\{\mathbf{H}\}$-space:

$$D_1 = \frac{(d^2 - 2)D^P - 2D^s}{d(d + 2)}, \quad D_2 = \frac{(d + 1)D^s + D^P}{d(d + 2)},$$

and d is the dimension of space.

Note that the one-point probability density is independent of variable $\mathbf{r}$ in the case of homogeneous initial conditions, and Eq. (9.21) assumes the form [28, 29, 36, 51]

$$\frac{\partial}{\partial t}P(t; \mathbf{H}) = \left\{D_1\frac{\partial^2}{\partial H_k \partial H_l}H_l H_k + D_2\frac{\partial^2}{\partial H_l \partial H_l}H_k^2\right\}P(t; \mathbf{H}).$$

(9.22)

Corollaries of Eq. (9.22) are the temporal behavior of average energy

$$\langle E(\mathbf{r}, t)\rangle = \langle \mathbf{H}^2(\mathbf{r}, t)\rangle$$

and the expression for the correlation of magnetic field components $\langle W_{ij}(\mathbf{r}, t)\rangle = \langle H_i(\mathbf{r}, t)H_j(\mathbf{r}, t)\rangle$

$$\langle E(\mathbf{r}, t)\rangle = E_0 \exp\left\{2\frac{d - 1}{d}(D^s + D^P)t\right\},$$

$$\frac{\langle W_{ij}(\mathbf{r}, t)\rangle}{\langle E(\mathbf{r}, t)\rangle} = \frac{1}{d}\delta_{ij} + \left(\frac{W_{ij}(0)}{E_0} - \frac{1}{d}\delta_{ij}\right)\exp\left\{-2\frac{(d + 1)D^s + D^P}{d + 2}t\right\}.$$

Thus, the mean magnetic field energy grows exponentially with time, and the isotropization of the magnetic field also develops exponentially. Note that the spectral components of the velocity field enter these exponents in an additive way. Obviously, this feature is preserved for all other correlations of the magnetic field and its energy.

9.2.2 Probabilistic Description of Magnetic Field Energy

To describe the magnetic field energy $E(\mathbf{r}, t) = \mathbf{H}^2(\mathbf{r}, t)$ in the Eulerian description, we introduce the indicator function:

$$\varphi(\mathbf{r}, t; E) = \delta(E(\mathbf{r}, t) - E),$$

in terms of which probability density of energy $P(\mathbf{r}, t; E)$ is defined as the equality

$$P(\mathbf{r}, t; E) = \langle \delta(E(\mathbf{r}, t) - E) \rangle_{\mathbf{u}} = \langle \delta(\mathbf{H}^2(\mathbf{r}, t) - E) \rangle_{\mathbf{H}}.$$

To derive an equation for this function, one should multiply Eq. (9.21) by function $\delta(\mathbf{H}^2 - E)$ and integrate the result over $\mathbf{H}$. The result will be the equation [28, 29, 36, 51]

$$\left(\frac{\partial}{\partial t} - D_0 \frac{\partial^2}{\partial \mathbf{r}^2} \right) P(\mathbf{r}, t; E) = \left\{ \alpha \frac{\partial}{\partial E} E + D \frac{\partial}{\partial E} E \frac{\partial}{\partial E} E \right\} P(\mathbf{r}, t; E),$$

$$P(\mathbf{r}, 0; E) = \delta(E - E_0(\mathbf{r}))$$

(9.23)

that coincides with Eq. (8.1) with the parameters

$$\alpha = 2 \frac{d - 1}{d + 2} \left(D^{\mathrm{P}} - D^{\mathrm{s}} \right), \quad D = 4(d - 1) \frac{(d + 1) D^{\mathrm{P}} + D^{\mathrm{s}}}{d(d + 2)}.$$

The parameter α can differ from zero (being positive or negative) or be equal to it (the critical case).

In the case of spatially homogeneous initial distribution of energy $E_0(\mathbf{r}) = E_0$, probability density is independent of $\mathbf{r}$ and is described by the formula

$$P(E, t; \alpha) = \frac{1}{2E\sqrt{\pi Dt}} \exp \left\{ -\frac{\ln^2 \left[Ee^{\alpha t} / E_0 \right]}{4Dt} \right\}.$$

(9.24)

Thus, in this case, the one-point statistical characteristics of energy $E(\mathbf{r}, t)/E_0$ are statistically equivalent to the characteristics of a random process $E(t)$ with probability density

$$P(E, t; \alpha) = \frac{1}{2E\sqrt{\pi Dt}} \exp \left\{ -\frac{\ln^2 \left(Ee^{\alpha t} \right)}{4Dt} \right\}.$$

(9.25)

The characteristic feature of distribution (9.25) lies in the appearance of a long sloping 'tail' for $Dt \gg 1$, implying an increased role of large outliers of the process $E(t)$ in forming its one-time statistics. For this distribution, all the moments of magnetic field energy, namely

$$\langle E^n(\mathbf{r}, t) \rangle = E_0^n \exp \left\{ -2n \frac{d-1}{d+2} \left(D^P - D^s \right) t + 4n^2 (d-1) \frac{(d+1) D^P + D^s}{d(d+2)} t \right\},$$

grow exponentially with time for $n > 0$, as well as for $n < 0$; in particular, at $n = 1$, the mean specific energy is written as

$$\langle E(\mathbf{r}, t) \rangle = E_0 \exp \left\{ \frac{2(d-1)}{d} (D^P + D^s) t \right\}, \tag{9.26}$$

or in a more suitable representation as

$$\left\langle \ln \frac{E(\mathbf{r}, t)}{E_0} \right\rangle = -\alpha t = -2 \frac{d-1}{d+2} \left(D^P - D^s \right) t.$$

The parameter α is, hence, the *Lyapunov characteristic index*. In this regard, the *typical realization curve* for the random process $E(t)$, determining the behavior of magnetic field energy in concrete realizations, is exponential at any fixed spatial location:

$$E^*(t) = E_0 e^{-\alpha t} = E_0 \exp \left\{ -2 \frac{d-1}{d+2} \left(D^P - D^s \right) t \right\},$$

and either grows or decays with time. So, for $\alpha > 0 \, (D^P > D^s)$ the typical realization curve decays exponentially at all spatial locations, which points to the cluster structure of a magnetic field in its individual realizations of a magnetic field energy; the growth of moments in this case is brought about by rare, yet strong, ejections of energy with respect to the typical realization curve.

For $\alpha < 0 \, (D^P < D^s)$ the typical realization curve grows exponentially with time, which points to a general increase in the magnetic field energy at each point in space. Clustering in the magnetic field energy does not happen in this case.

Note that namely this case is realized for an incompressible magnetohydrodynamic flow ($D^P = 0$); hence, *there is no structure formation in the incompressible case*.

The respective asymptotic expressions for the specific values of the volume of large ejections and their total energy, for homogeneous initial conditions, have the form of expressions (8.13) and (8.14) with the change $f \to E$.

This implies that for $\alpha > 0 \, (D^P > D^s)$ the specific total volume tends to zero, while the specific total energy inside it coincides with the mean energy in the entire space. The latter points to clustering of magnetic field energy with a probability one, i.e., in almost all magnetic field realizations. Consequently, the magnetic field is simply absent over a larger portion of the space.

When $\alpha < 0$ $(D^P < D^s)$, clustering does not happen, and there is only a general increase in the magnetic field energy everywhere in space. Note that, in this case, the inverse quantity, $1/E$, undergoes clustering, i.e., clusters of compact regions appear from where the magnetic field is expelled (magnetic zeros).

We mentioned earlier that the parameters D^P and D^s, characterizing statistics of the random velocity field enter all statistical moment and correlation functions of the magnetic field energy in an additive way. This is, certainly, the consequence of Eqs. (6.46) and (6.52) being linear. However, this fact implies that all the main (functional) relationships in such a statistical description do not *distinguish between the influence of solenoidal and potential components* of the random velocity field. This means that all the relationships derived for the mentioned statistical quantities have the same structure for both incompressible $(D^P = 0)$, and purely potential $(D^s = 0)$ flows. Nevertheless, since clustering is absent for an incompressible flow and, conversely, is present for a potential one, it is absolutely clear that the statistical characteristics mentioned above *do not contain any information on the stochastic structure formation in individual realizations of magnetic field energy, namely, on clustering*.

Furthermore, the input induction equation (6.46) is valid within the applicability limits of the kinematic approximation. In the presence of clustering, when the magnetic field is absent over a larger portion of space, its back reaction on the velocity field is not essential. In contrast, in the absence of clustering, when the magnetic field is generated everywhere in space, the kinematic approximation can be valid only over a small time interval, and discussing the role of the dynamic diffusion coefficient in forming the statistics of magnetic field energy in this interval is, in our opinion, simply not serious.

9.2.3 The Critical Case of $\alpha = 0$ $(D^P = D^s)$

This case can be considered as *pseudoequilibrium*, in analogy with the case of equilibrium thermal noise [68, 69]. At $\alpha = 0$, the one-point probability density takes the form

$$P(t; E) = \frac{1}{2E\sqrt{\pi Dt}} \exp\left\{-\frac{\ln^2[E/E_0]}{4Dt}\right\}.$$

The random processes $E(t)$ and $1/E(t)$ are statistically equivalent. The specific mean volume tends to a half of the total volume as $t \to \infty$, and the specific mean energy tends to the total mean energy.

Thus, clustering does not happen at $\alpha = 0$ $(D^P = D^s)$ in the framework of Eq. (8.4). It is worth mentioning that this result does not seem satisfactory, for Eq. (8.4) is itself approximate, obtained under the assumption that the random velocity field is delta-correlated in time.

Accounting for the finiteness of the time correlation radius allows unequivocally judging the presence or lack of the physical phenomenon of *clustering*. The results of dedicated computations have shown that accounting for the finite time correlation radius of a velocity field clustering occurs with probability one [69].

An analogous situation also takes place for random acoustic waves if they are not damped.

Thus, a random Gaussian acoustic field $\mathbf{u}(\mathbf{r}, t)$, statistically homogeneous and isotropic in space and stationary in time, is described by the correlation and spectral tensors $(\tau = t - t')$

$$\langle u_i(\mathbf{r}, t) u_j(\mathbf{r}', t') \rangle = \sigma_{\mathbf{u}}^2 B_{ij}(\mathbf{r} - \mathbf{r}', \tau) = \sigma_{\mathbf{u}}^2 \int d\mathbf{k} \, E_{ij}(\mathbf{k}) f(\mathbf{k}, \mathbf{r}, \tau),$$

where $\sigma_{\mathbf{u}}^2 = \langle \mathbf{u}^2(\mathbf{r}, t) \rangle$ is the variance of the velocity field, and the function

$$f(\mathbf{k}, \mathbf{r}, \tau) = e^{-\lambda(k)\tau} \cos\{\mathbf{kr} - \omega(k)\tau\},$$

where $\omega(k) = ck$ is the dispersion curve for acoustic waves, and c is the speed of sound.

The exponentially decaying term is related to dissipative factors in equations of fluid dynamics and magnetohydrodynamics, and $\lambda(k) = \lambda_p k^2$.

Here, the velocity field spectral tensor contains only the potential component $E_{ij}(\mathbf{k}) = E(k) \dfrac{k_i k_j}{k^2}$. And, since the time integral for $\lambda_p \ll cl_0$ (where l_0 is the spatial correlation radius of the velocity field) has the asymptotics

$$\int\limits_0^\infty dt \, f(k, t) = \lambda_p/c^2,$$

clustering of the magnetic field energy occurs with a probability one (i.e., in almost all its realizations) in the presence of small absorption.

In the absence of dissipation, the diffusion coefficient in Eq. (8.4) becomes zero, and we do not have any information on the presence or absence of *clustering*. In general, however, the following equality is valid in the absence of dissipation:

$$\int\limits_0^\infty dt \, \cos\{\omega(k)t\} = \pi\delta\left(\omega(k)\right),$$

owing to which resonances emerge between various harmonics of the acoustic wave field in higher approximations. This allows establishing in the second order of the functional perturbation method (after cumbersome manipulations) that clustering of magnetic field energy occurs with a unit probability (i.e., in almost all its realizations), and computing the characteristic time it takes for clustering to set up, $t \sim 1/\alpha_2$, where the Lyapunov characteristic parameter [69] is given by

$$\alpha_2 = \frac{\sigma_u^2}{c^2} \int d\mathbf{k} \; k^2 E(k) \left[\frac{4}{5} \lambda_P + 76\pi^2 \frac{\sigma_u^2}{c} k^2 E(k) \right].$$

Notice that there is no clustering of magnetic field energy in an equilibrium thermal velocity field.

We also mention that analogous computations for the density field of the passive tracer in random wave fields have indicated the presence of clustering with a probability one [70].

Chapter 10
Parametrically Excited Dynamic Systems with Gaussian Pumping

10.1 Statistical Analysis of Simple Turbulent Dynamo Problem with Gaussian Pumping

Turn now to the simplest model of the velocity field (3.4) that allows a solution in the analytic form (3.55). The equation for the one-point in space and one-time probability density of magnetic field energy, written for dimensionless time (3.11) $\tau = k^2 \sigma^2 \tau_0 t$ (see Refs. [51, 72] and monograph [28, 29]) which corresponds to linear Eq. (3.54), is written as

$$\frac{\partial}{\partial \tau} P(\tau; E) = \left(\frac{\partial}{\partial E} E + 2 \frac{\partial}{\partial E} E \frac{\partial}{\partial E} E + 2 \frac{\partial}{\partial E} E \frac{\partial}{\partial E} \right) P(\tau; E) \qquad (10.1)$$

with the initial condition $P(0; E) = \delta(E - \beta)$.

For the last equation, the large-time asymptotics for the moments of magnetic field energy was obtained in the form

$$\langle E^n(\tau) \rangle \sim A_n e^{n(2n-1)\tau},$$

which corresponds to the lognormal law for the probability density with a correction that accounts for the Gaussian field generation. Also, the expression for the Lyapunov exponent

$$e^{\langle \ln E(\tau) \rangle} = \beta e^{-\tau}$$

was derived, which indicates that the magnetic field energy decays at any point in space, i.e., that the clustering proceeds.

The last term in Eq. (10.1) is responsible for the generation of a Gaussian magnetic field, which dominates the magnetic field energy generation at small times. We present below the related equation and solutions for these times, bounded to the case of two dimensions which is of interest to us [83].

© Springer International Publishing AG 2017
V.I. Klyatskin, *Fundamentals of Stochastic Nature Sciences*,
Understanding Complex Systems, DOI 10.1007/978-3-319-56922-2_10

The equation for the probability density of the two-dimensional Gaussian vector field $\mathbf{H}_\perp(\mathbf{R}, \mathbf{t})$ for the spatially homogeneous case takes the form

$$\frac{\partial}{\partial \tau} P(\tau; \mathbf{H}_\perp) = \frac{1}{2} \frac{\partial^2}{\partial \mathbf{H}_\perp^2} P(\tau; \mathbf{H}_\perp),$$

and its solution is written as

$$P(\tau; \mathbf{H}_\perp) = \frac{1}{2\pi\tau} \exp\left(-\frac{\mathbf{H}_\perp^2}{2\tau}\right).$$

Respectively, on this short time interval, the probability density for the transverse energy $E = \mathbf{H}_\perp^2(\mathbf{R}, t)$, defined as

$$P(\tau; E) = \frac{1}{2\tau} \exp\left(-\frac{E}{2\tau}\right) \tag{10.2}$$

is described by the equation

$$\frac{\partial}{\partial \tau} P(\tau; E) = 2\frac{\partial}{\partial E} E \frac{\partial}{\partial E} P(\tau; E).$$

As a consequence, the integral probability distribution function is governed by the equation

$$\frac{\partial}{\partial \tau} \Phi(\tau; E) = 2E \frac{\partial^2}{\partial E^2} \Phi(\tau; E),$$

the solution of which is written out as

$$\Phi(\tau; E) = 1 - \exp\left(-\frac{E}{2\tau}\right). \tag{10.3}$$

This relationship leads directly to the following expression for the typical realization curve: $E^*(\tau) = (2 \ln 2)\tau$.

Clearly, for clustering to occur, the typical realization curve of a respective process must decay, in contrast to linear growth for a Gaussian process. From Fig. 10.1, which displays the results of a numerical solution of Eq. (10.1), it can be seen that at the initial stage the probability density of the magnetic field energy decays approximately as the Gaussian distribution, and the decay rate decreases with time, i.e., the process of Gaussian field generation prevails. At times of about $\tau = 1.7$, the situation changes: at larger times, clustering begins to play an important role, i.e., the rate of decrease in probability density begins to increase with time as the energy increases, just like its value at zero.

Consider now the integral probability function for the magnetic field energy. From Eq. (10.1), we obtain in a regular way the equation for its evolution:

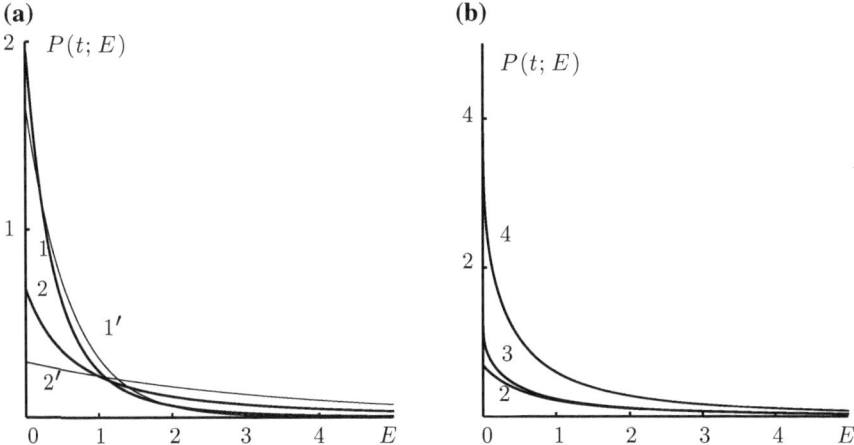

Fig. 10.1 Probability density (10.1) for the time: **a** $\tau = 0.3$ (*curve 1*), $\tau = 1.7$ (*curve 2*). The *thin lines* correspond to the Gaussian distribution (10.2) for $\tau = 0.3$ (*curve 1'*) and $\tau = 1.7$ (*curve 2'*). **b** The same as in panel (**a**), but for the time: $\tau = 1.7$ (*curve 2*), $\tau = 5.0$ (*curve 3*), and $\tau = 8.3$ (*curve 4*)

$$\frac{\partial}{\partial \tau} \Phi(\tau; E) = \left(E + 2E \frac{\partial}{\partial E} (E + 1) \right) \frac{\partial}{\partial E} \Phi(\tau; E). \tag{10.4}$$

A numerical solution to the last equation is given in Fig. 10.2a; it indicates that at relatively small times, approximately until $\tau = 1.7$, the rate of increase in the integral function decreases with time, which is characteristic of the Gaussian distribution. At larger times, the rate of increasing begins to increase, which is characteristic of the lognormal distribution.

This observation is illustrated more transparently by the typical realization curve. We see in Fig. 10.2b that for $\tau \geq 3.0$ the typical realization curve decays for the process described by Eq. (10.1), indicating the presence of clustering. It is noteworthy that the process of clustering manifests itself even earlier, but it prevails over the generation from approximately this time instant.

Figure 10.3a plots the time dependence of the specific area of the regions where the magnetic field energy exceeds the maximum level of the typical realization curve, i.e., $E > 2$. In the case of the Gaussian distribution, this area tends to unity, which testifies to a lack of clustering; however, in the presence of a magnetic field, the area begins to decrease, which points to the beginning of clustering at times $\tau \sim 1.7$. Figure 10.3b shows the temporal dynamics of the specific energy (normalized to the total energy at the respective time instant) confined in these regions, thus confirming the phenomenon of clustering.

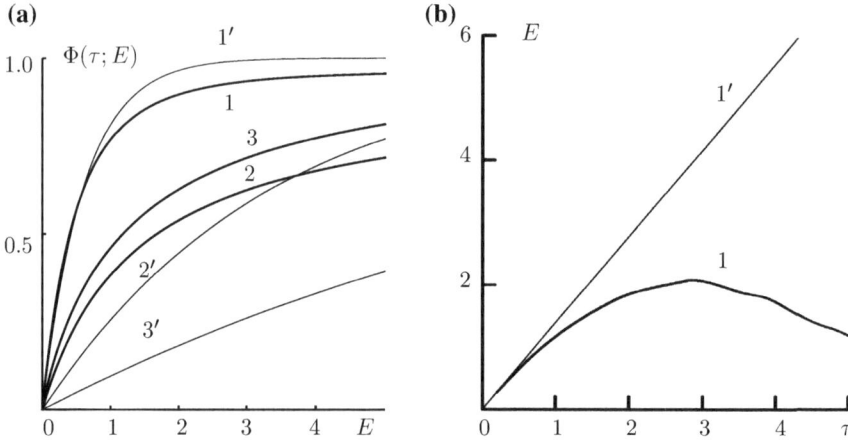

Fig. 10.2 **a** Integral probability function (10.4) for three moments of dimensionless time: $\tau = 0.3$ (*curve 1*), $\tau = 1.7$ (*curve 2*), $\tau = 5.0$ (*curve 3*); The *thin curves* $1'-3'$ correspond to the Gaussian distribution for the same time instants, **b** Typical realization curves for magnetic field energy (1) and a Gaussian process ($1'$)

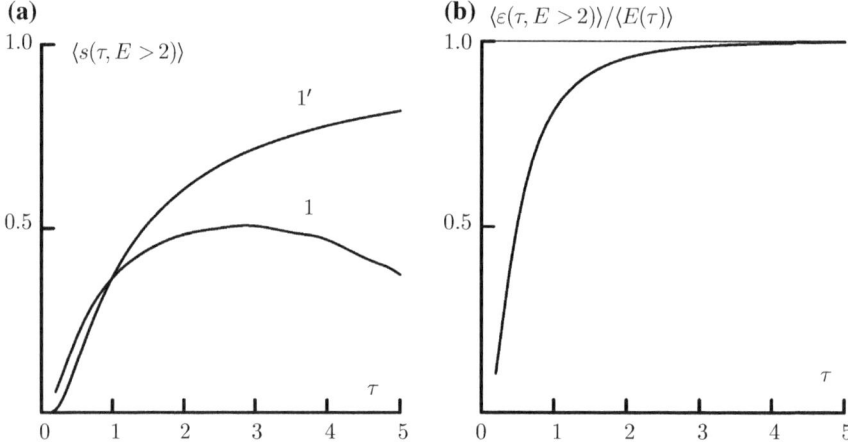

Fig. 10.3 **a** Specific area of regions for distribution (10.1), where the level of magnetic field energy $E > 2$. **b** Specific energy confined in regions whose evolution is shown in panel (**a**)

10.2 Anomalous Sea Surface Structures

Recent decades have seen an increased interest in such phenomena as anomalously large waves, also called rogue or freak waves (see, for example, Ref. [73]). There are many ideas on the mechanisms and methods to describe this phenomenon [73–75]. We presume that different mechanisms may exist, and not all of them deserve to be called anomalously high waves. For example, the development of such waves against the background of sufficiently high sea waves is apparently related to nonlinear effects (see, for example, Refs. [76–79], and references cited therein). Different dynamic models are being studied based on numerical modeling, as well as on analytical results related to the nonlinear Schródinger equation. The statistical properties of observed anomalous waves are also intensely being discussed (see, for example, the review [74]).

We propose a possible mechanism for the emergence of anomalous structures on the water surface, which may correspond to the appearance of such structures against the background of very weak waviness. Figures 10.4 and 10.5 present three photos of an unusually narrow and elongated structure 4–5 m in height observed on 11 June 2006 in the vicinity of the Kamchatka Pacific coast, 1–1.5 km offshore [29, 82]. The photographer, Sokolovsky, describes the phenomenon documented by him in the following way: "It was, surely, a strange wave, for it was repeated several times, each time disappearing. There were no waves around, a completely still surface".

Figure 10.6 also demonstrates the structure of the sea surface observed close to the coast of Île de Ré (France) [80].[1]

From these examples, it is clear that if we construct a topographic map of the magnitude of the sea surface elevation gradient for them, we obtain a typical cluster structure for which a positive field is concentrated within a small area, being simply absent elsewhere.

It should be emphasized once again that the structures discussed above are notably different from the anomalous waves commonly considered. First, such structures can be both standing and moving, and second, they substantially exceed the background; here, we are not dealing with outliers that exceed the background just twofold [74]. In the examples above, one may rather speculate about the presence of high structures against a practically vanishing background. From these examples, it is also seen that the clustering of the positive field in this case is equivalent to such physical phenomena as the focusing of a wave field on passing through a random medium [74, 81] or of the intensity of laser radiation traversing a random medium (see Sect. 8.2.1).

10.2.1 Problem Statement

The statement of the statistical problem on the emergence of anomalous structures on the sea surface in the kinematic approximation is presented in Ref. [82] and monograph [29].

We denote the three-dimensional spatial coordinate as $\mathbf{r} = \{r_i\}$, where $i = 1, 2, 3, z = r_3$ being the vertical coordinate as, and use R_α ($\alpha = 1, 2$) for the coordinates in the horizontal plane perpendicular to the z-axis. In this notation, $\mathbf{r} = \{\mathbf{R}, z\}$. Accordingly, we represent the three-dimensional hydrodynamic velocity

---

[1]Impressive photos of other anomalous structures on the sea surface can be found at the site http://imgur.eom/a/4Y2Oo.

Fig. 10.4   View on the water structure from the side

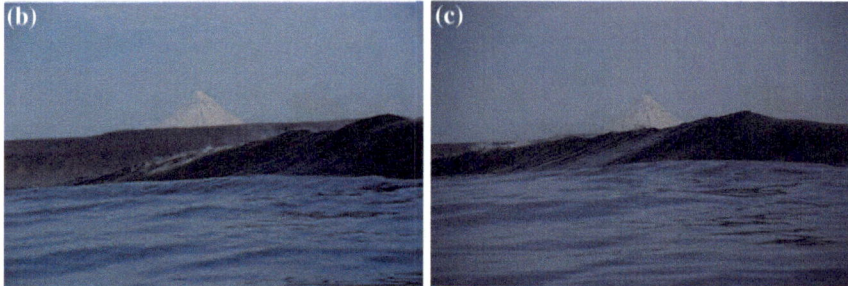

Fig. 10.5   The front view: the beginning (**a**), and the middle (**b**) of the structure

Fig. 10.6   Example of a water structure on the sea surface

Fig. 10.7 Perturbation of
the sea surface

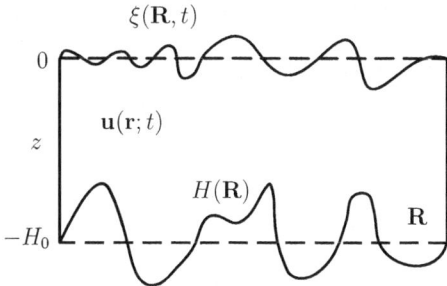

field $\mathbf{u}(\mathbf{r}; t)$ through its horizontal and vertical components, i.e., in the form $u_i(\mathbf{r}; t) = \{u_\alpha(\mathbf{R}, z; t), \ w(\mathbf{R}, z; t)\}$, where the subscripts $i = \{1, \ 2, \ 3\}$ and $\alpha = \{1, \ 2\}$.

The sea surface elevation (displacement) $z = \xi(\mathbf{R}, t)$ is described through the kinematic boundary condition (Fig. 10.7), which is expressed as

$$\frac{d}{dt}\xi(\mathbf{R}, t) = w_z(\mathbf{R}, z; t)\Big|_{z=\xi(\mathbf{R},t)}. \tag{10.5}$$

Here, $\dfrac{d}{dt}\xi(\mathbf{R}, t)$ is the total derivative of a sea surface elevation.

Boundary condition (10.5) can be considered a closed stochastic quasilinear equation in the framework of the kinematic approximation. Namely, for the given statistical characteristics of the velocity fields $\mathbf{u}(\mathbf{R}, z; t)$ and $w(\mathbf{R}, z; t)$ this equation has the form

$$\frac{\partial \xi(\mathbf{R}, t)}{\partial t} + u_\alpha(\mathbf{R}, \xi(\mathbf{R}, t), t)\frac{\partial \xi(\mathbf{R}, t)}{\partial R_\alpha} = w_z(\mathbf{R}, \xi(\mathbf{R}, t); t) \tag{10.6}$$

with the initial condition $\xi(\mathbf{R}, 0) = \xi_0(\mathbf{R})$. Equation (10.6) describes the generation of waves on the sea surface excited by a Gaussian vertical component of a hydrodynamic velocity field.

Differentiating Eq. (10.6) with respect to $\mathbf{R}$, we arrive at the equation for the gradient of sea surface elevation $p_\beta(\mathbf{R}, t) = \dfrac{\partial \xi(\mathbf{R}, t)}{\partial R_\beta}$ characterizing the slope of the sea surface:

$$\frac{\partial p_\beta(\mathbf{R}, t)}{\partial t} + \left[\frac{\partial u_\alpha(\mathbf{R}, z; t)}{\partial R_\beta} + \frac{\partial u_\alpha(\mathbf{R}, z; t)}{\partial z}p_\beta(\mathbf{R}, t)\right]_{z=\xi(\mathbf{R},t)} p_\alpha(\mathbf{R}, t)$$

$$+ u_\alpha(\mathbf{R}, \xi(\mathbf{R}, t), t)\frac{\partial p_\alpha(\mathbf{R}, t)}{\partial R_\beta}$$

$$= \left[\frac{\partial w(\mathbf{R}, z; t)}{\partial R_\beta} + \frac{\partial w(\mathbf{R}, z; t)}{\partial z}p_\beta(\mathbf{R}, t)\right]_{z=\xi(\mathbf{R},t)} p_\alpha(\mathbf{R}, t), \tag{10.7}$$

with the boundary condition $\mathbf{p}(\mathbf{R}, 0) = \mathbf{p}_0(\mathbf{R}) = \dfrac{\partial \xi_0(\mathbf{R})}{\partial \mathbf{R}}$.

Notice that one more boundary condition, related to the inhomogeneities in bottom topography (see Fig. 10.7), exists for the problem considered. In the framework of the kinematic approximation, this boundary condition is manifested in a functional form, namely, for variational derivatives of problem solutions $\xi(\mathbf{R}, t)$ and $\mathbf{p}(\mathbf{R}, t)$ the following relationships are valid:

$$
\begin{aligned}
\frac{\delta \xi(\mathbf{R}, t)}{\delta \mathbf{u}(\mathbf{R}', z', t')} &\sim \theta \left(z' + H_0 - H(\mathbf{R})\right) \theta \left(t - t'\right), \\
\frac{\delta \mathbf{p}(\mathbf{R}, t)}{\delta \mathbf{u}(\mathbf{R}', z', t')} &\sim \theta \left(z' + H_0 - H(\mathbf{R})\right) \theta \left(t - t'\right),
\end{aligned}
\tag{10.8}
$$

where $\theta(z)$ is the Heaviside theta-function. Condition (10.8) corresponds to an impermeable sea bottom.

Thus, Eqs. (10.6) and (10.7) together with boundary condition (10.8) represent a closed system in the kinematic approximation. The solution of this system should provide an answer to the question of whether or not its equations contain information on the presence of anomalous structures on the sea surface with a probability one, i.e., for almost any realization of the random velocity field. But we put aside here the physical mechanisms underlying this structure formation.

Let us introduce a joint indicator function of surface elevation and its gradient,

$$
\varphi(\mathbf{R}, t; \xi, \mathbf{p}) = \delta \left(\xi(\mathbf{R}, t) - \xi\right) \delta \left(\mathbf{p}(\mathbf{R}, t) - \mathbf{p}\right).
\tag{10.9}
$$

Taking into account dynamic conditions, one can derive a linear Liouville equation [29, 82]

$$
\begin{aligned}
\frac{\partial \varphi(\mathbf{R}, t; \xi, \mathbf{p})}{\partial t} = &-\frac{\partial}{\partial \xi} w(\mathbf{R}, \xi; t) \varphi(\mathbf{R}, t; \xi, \mathbf{p}) \\
&- \left[u_\alpha(\mathbf{R}, \xi; t) \frac{\partial}{\partial R_\alpha} - \frac{\partial u_\alpha(\mathbf{R}, \xi; t)}{\partial \xi} p_\alpha \right] \varphi(\mathbf{R}, t; \xi, \mathbf{p}) \\
&- \frac{\partial}{\partial p_\beta} \left[\frac{\partial u_\alpha(\mathbf{R}, \xi; t)}{\partial R_\beta} + \frac{\partial u_\alpha(\mathbf{R}, \xi; t)}{\partial \xi} p_\beta \right] p_\alpha \varphi(\mathbf{R}, t; \xi, \mathbf{p}) \\
&- \frac{\partial}{\partial p_\beta} \left[\frac{\partial w(\mathbf{R}, \xi; t)}{\partial R_\beta} + \frac{\partial w(\mathbf{R}, \xi; t)}{\partial \xi} p_\beta \right] \varphi(\mathbf{R}, t; \xi, \mathbf{p})
\end{aligned}
\tag{10.10}
$$

with initial conditions

$$
\varphi(\mathbf{R}, 0; \xi, \mathbf{p}) = \delta \left(\xi - \xi_0(\mathbf{R})\right) \delta(\mathbf{p} - \mathbf{p}_0(\mathbf{R})).
$$

Note that (10.10) describes the joint probability density of surface displacement and its spatial gradient for dynamic systems with deterministic parameters under random initial conditions.

Correlation functions of fields $u_\alpha(\mathbf{R}, \xi; t)$ and $w(\mathbf{R}, \xi; t)$

The required statistical characteristics follow from statistical properties of the components $u_\alpha(\mathbf{R}, \xi; t)$ and $w(\mathbf{R}, \xi; t)$ of the hydrodynamics field $\mathbf{u}(\mathbf{r}, t)$, in the Liouville equation (10.10).

Correlation functions of interest here can be obtained from Eq. (4.2) rewritten in the two-dimensional vector $\mathbf{K}$ and one-dimensional projection $z = \xi$. The resulting formula is as follows,

$$B_{\alpha\beta}(\mathbf{R} - \mathbf{R}', \xi - \xi', t - t') = \langle u_\alpha(\mathbf{R}, \xi; t) u_\beta(\mathbf{R}', \xi'; t') \rangle$$

$$= \int d\mathbf{k} \left[E^s(k, t - t') \left(\delta_{\alpha\beta} - \frac{K_\alpha K_\beta}{k^2} \right) + E^P(k, t - t') \frac{K_\alpha K_\beta}{k^2} \right]$$

$$\times e^{i\mathbf{K}(\mathbf{R}-\mathbf{R}')+ik_z(\xi-\xi')}, \quad (10.11)$$

where $\mathbf{k} = \{\mathbf{K}, k_z\}$, $\mathbf{K} = \{K_a\}$, $\alpha = 1, 2$, and $k = \sqrt{\mathbf{K}^2 + k_z^2}$.

Similar formulas were derived for correlations $B_{\alpha w}$ and B_{ww}

$$B_{\alpha w}(\mathbf{R} - \mathbf{R}', \xi - \xi'; t - t') = -\int d\mathbf{k} \left[E^s(k, t - t') - E^P(k, t - t') \right] \frac{K_\alpha k_z}{k^2}$$

$$\times e^{i\mathbf{K}(\mathbf{R}-\mathbf{R}')+ik_z(\xi-\xi')}, \quad (10.12)$$

$$B_{ww}(\mathbf{R} - \mathbf{R}', \xi - \xi'; t - t') = \int d\mathbf{k} \left[E^s(k, t - t') \left(1 - \frac{k_z^2}{k^2} \right) + E^P(k, t - t') \frac{k_z^2}{k^2} \right]$$

$$\times e^{i\mathbf{K}(\mathbf{R}-\mathbf{R}')+ik_z(\xi-\xi')}. \quad (10.13)$$

Correspondingly, the second derivative of correlation function $B_{\alpha\beta}(\mathbf{R}, \xi)$ (10.11) at zero-valued argument has the form

$$\frac{\partial^2 B_{\alpha\beta}(0, 0)}{\partial R_\gamma \partial R_\delta} = -\int d\mathbf{k} \, K_\gamma K_\delta \left[E^s(k) \left(\delta_{\alpha\beta} - \frac{K_\alpha K_\beta}{k^2} \right) + E^P(k) \frac{K_\alpha K_\beta}{k^2} \right],$$

$$(10.14)$$

where $E^s(k) = \int_0^\infty d\tau E^s(k, \tau)$, $E^P(k) = \int_0^\infty d\tau E^P(k, \tau)$, and we have for other correlation functions the expressions

$$\frac{\partial^2 B_{\alpha w}(\mathbf{0}, 0)}{\partial R_\alpha \partial \xi} = \int d\mathbf{k} \, \frac{K^2 k_z^2}{k^2} \left[E^s(k) - E^P(k) \right],$$

$$\frac{\partial^2 B_{\alpha\alpha}(\mathbf{0}, 0)}{\partial \xi \partial \xi} = - \int d\mathbf{k} \left[E^s(k) \left(2k_z^2 - \frac{K^2 k_z^2}{k^2} \right) + E^P(k) \frac{K^2 k_z^2}{k^2} \right],$$

$$\frac{\partial^2 B_{ww}(\mathbf{0}, 0)}{\partial R_\gamma \partial R_\gamma} = - \int d\mathbf{k} \left[E^s(k) \left(K^2 - \frac{K^2 k_z^2}{k^2} \right) + E^P(k) \frac{K^2 k_z^2}{k^2} \right],$$

$$\frac{\partial^2 B_{ww}(\mathbf{0}, 0)}{\partial \xi \partial \xi} = - \int d\mathbf{k} \left[E^s(k) \left(k_z^2 - \frac{k_z^4}{k^2} \right) + E^P(k) \frac{k_z^4}{k^2} \right].$$

$$(10.15)$$

It is worth recalling that all the statistical characteristics of the velocity field components relate to the statistical characteristics of the field $\mathbf{u}(\mathbf{r}, t)$. The relation is clearly seen in the polar coordinates $k_z = k \cos \theta$, $K^2 = k^2 \sin^2 \theta$, after integration over the angle variables.

10.2.2 Equation in Probability Density

The case considered imply that the joint probability density of surface elevation and its gradient is the indicator function (10.9) averaged over an ensemble of the random velocity field realizations $\mathbf{u}(\mathbf{r}, t)$, i.e.

$$P(\mathbf{R}, t; \xi, \mathbf{p}) = \langle \varphi(\mathbf{R}, t; \xi, \mathbf{p}) \rangle_{\mathbf{u}}. \tag{10.16}$$

Now one can obtain an equation governing the joint probability density of surface elevation and its gradient. To do that, one averages Liouville equation (10.10) over an ensemble of the random velocity field realizations $\mathbf{u}(\mathbf{r}, t)$. To separate velocity correlations $\mathbf{u}(\mathbf{r}, t)$, one can make use of the Furutsu-Novikov formula (see, for example, monographs [28, 29] and Appendix), accounting for sea bottom irregularities $H(\mathbf{R})$ (see Fig. 10.7 and Eq. (10.8)):

$$\langle u_i(\mathbf{R}, \xi, t) R \left[\mathbf{u}(\tilde{\mathbf{R}}, \tilde{\xi}, \tau) \right] \rangle_{\mathbf{u}} = \int d\mathbf{R}' \int_{-\infty}^{\xi+0} \theta \left(\xi' + H_0 - H(\mathbf{R}) \right) d\xi'$$

$$\times \int_0^t dt' \, B_{ij}(\mathbf{R} - \mathbf{R}', \xi - \xi', t - t') \left\langle \frac{\delta R \left[\mathbf{u}(\tilde{\mathbf{R}}, \tilde{z}, \tau) \right]}{\delta u_j(\mathbf{R}', \xi', t')} \right\rangle_{\mathbf{u}}. \tag{10.17}$$

Assuming bottom topography specified statistically with zero mean value, we average Eq. (10.17) over an ensemble of its realizations. Under the assumption that topographic inhomogeneties of bottom are statistically independent of the field of hydrodynamic velocities, we obtain the following expression

$$\left\langle u_i(\mathbf{R}, \xi, t) R\left[\mathbf{u}(\widetilde{\mathbf{R}}, \widetilde{\xi}, \tau)\right]\right\rangle_{\mathbf{u}} = \int d\mathbf{R}' \int_{-\infty}^{\xi+0} \left\langle \theta\left(\xi' + H_0 - H(\mathbf{R})\right)\right\rangle_H d\xi'$$

$$\times \int_0^t dt' B_{ij}(\mathbf{R} - \mathbf{R}', \xi - \xi', t - t') \left\langle \frac{\delta R\left[\mathbf{u}(\widetilde{\mathbf{R}}, \widetilde{z}, \tau)\right]}{\delta u_j(\mathbf{R}', \xi', t')} \right\rangle_{\mathbf{u}}, \quad (10.18)$$

where function $\left\langle \theta\left(\xi + H_0 - H(\mathbf{R})\right)\right\rangle_H$ is the integral probability distribution of topographic inhomogeneities $H(\mathbf{R})$, i.e., the probability density $\mathsf{P}\{\xi > -H_0 + H(\mathbf{R})\}$. In the case of statistically homogeneous random field $H(\mathbf{R})$, this function is independent of spatial point $\mathbf{R}$, i.e., $\left\langle \theta\left(\xi + H_0 - H(\mathbf{R})\right)\right\rangle = \mathsf{P}(H_0; \xi)$.

For an infinitely deep sea (at $H_0 \to \infty$ in Fig. 10.7) we have $\mathsf{P}(H_0; \xi) \to 1$.

In the general case the use of the diffusion approximation simplifies Eq. (10.18) and reduces it to the form

$$\left\langle u_i(\mathbf{R}, \xi, t) R\left[\mathbf{u}(\widetilde{\mathbf{R}}, \widetilde{\xi}, \tau)\right]\right\rangle_{\mathbf{u}} = \int d\mathbf{R}' \int_{-\infty}^{\xi+0} \mathsf{P}(H_0; \xi') d\xi'$$

$$\times B_{ij}(\mathbf{R} - \mathbf{R}', \xi - \xi') \left\langle \frac{\delta R\left[\mathbf{u}(\widetilde{\mathbf{R}}, \widetilde{z}, \tau)\right]}{\delta u_j(\mathbf{R}', \xi', t - 0)} \right\rangle_{\mathbf{u}},$$

where function $B_{ij}(\mathbf{r})$ is described by Eq. (4.4), i.e.

$$B_{ij}(\mathbf{R}, \xi) = \int_0^\infty d\tau \, B_{ij}(\mathbf{R}, \xi, \tau). \quad (10.19)$$

We have, in particular,

$$\left\langle u_\alpha(\mathbf{R}, \xi, t) \varphi\left[\mathbf{u}(\widetilde{\mathbf{R}}, \widetilde{\xi}, \tau)\right]\right\rangle_{\mathbf{u}}$$

$$= \int d\mathbf{R}' \int_{-\infty}^{\xi+0} \mathsf{P}(H_0; \xi') d\xi' \, B_{\alpha\beta}(\mathbf{R} - \mathbf{R}', \xi - \xi') \left\langle \frac{\delta \varphi\left[\mathbf{u}(\widetilde{\mathbf{R}}, \widetilde{z}, \tau)\right]}{\delta u_\beta(\mathbf{R}', \xi', t - 0)} \right\rangle_{\mathbf{u}}$$

$$+ \int d\mathbf{R}' \int_{-\infty}^{\xi+0} \mathsf{P}(H_0; \xi') d\xi' \, B_{\alpha w}(\mathbf{R} - \mathbf{R}', \xi - \xi') \left\langle \frac{\delta \varphi\left[\mathbf{u}(\widetilde{\mathbf{R}}, \widetilde{z}, \tau)\right]}{\delta w(\mathbf{R}', \xi', t - 0)} \right\rangle_{\mathbf{u}},$$

$$\left\langle w(\mathbf{R}, \xi, t) \varphi\left[\mathbf{u}(\widetilde{\mathbf{R}}, \widetilde{\xi}, \tau)\right]\right\rangle_{\mathbf{u}}$$

$$= \int d\mathbf{R}' \int_{-\infty}^{\xi+0} \mathsf{P}(H_0; \xi') d\xi' \, B_{w\beta}(\mathbf{R} - \mathbf{R}', \xi - \xi') \left\langle \frac{\delta \varphi\left[\mathbf{u}(\widetilde{\mathbf{R}}, \widetilde{z}, \tau)\right]}{\delta u_\beta(\mathbf{R}', \xi', t - 0)} \right\rangle_{\mathbf{u}}$$

$$+ \int d\mathbf{R}' \int\limits_{-\infty}^{\xi+0} d\xi' \mathsf{P}(H_0; \xi') \, B_{ww}(\mathbf{R} - \mathbf{R}', \xi - \xi') \left\langle \frac{\delta \varphi\left[\mathbf{u}(\widetilde{\mathbf{R}}, \widetilde{z}, \tau)\right]}{\delta w(\mathbf{R}', \xi', t - 0)} \right\rangle_{\mathbf{u}}.$$

Thus, averaging the Liouville equation (10.10) over an ensemble of velocity field realizations, we obtain an equation in probability density of the form

$$\frac{\partial P(\mathbf{R}, t; \xi, \mathbf{p})}{\partial t}$$

$$= - \int d\mathbf{R}' \int\limits_{-\infty}^{\xi+0} \mathsf{P}(H_0; \xi') d\xi' \, B_{\alpha\beta}(\mathbf{R} - \mathbf{R}', \xi - \xi') \frac{\partial}{\partial R_\alpha} \left\langle \frac{\delta \varphi\left[\mathbf{u}(\widetilde{\mathbf{R}}, \widetilde{z}, \tau)\right]}{\delta u_\beta(\mathbf{R}', \xi', t - 0)} \right\rangle$$

$$- \int d\mathbf{R}' \int\limits_{-\infty}^{\xi+0} \mathsf{P}(H_0; \xi') d\xi' \, B_{\alpha w}(\mathbf{R} - \mathbf{R}', \xi - \xi') \frac{\partial}{\partial R_\alpha} \left\langle \frac{\delta \varphi\left[\mathbf{u}(\widetilde{\mathbf{R}}, \widetilde{z}, \tau)\right]}{\delta w(\mathbf{R}', \xi', t - 0)} \right\rangle$$

$$+ \int d\mathbf{R}' \int\limits_{-\infty}^{\xi+0} \mathsf{P}(H_0; \xi') d\xi' \, \frac{\partial B_{\alpha\beta}(\mathbf{R} - \mathbf{R}', \xi - \xi')}{\partial \xi} p_\alpha \left\langle \frac{\delta \varphi\left[\mathbf{u}(\widetilde{\mathbf{R}}, \widetilde{z}, \tau)\right]}{\delta u_\beta(\mathbf{R}', \xi', t - 0)} \right\rangle$$

$$+ \int d\mathbf{R}' \int\limits_{-\infty}^{\xi+0} \mathsf{P}(H_0; \xi') d\xi' \, \frac{\partial B_{\alpha w}(\mathbf{R} - \mathbf{R}', \xi - \xi')}{\partial \xi} p_\alpha \left\langle \frac{\delta \varphi\left[\mathbf{u}(\widetilde{\mathbf{R}}, \widetilde{z}, \tau)\right]}{\delta w(\mathbf{R}', \xi', t - 0)} \right\rangle$$

$$- \frac{\partial}{\partial \xi} \int d\mathbf{R}' \int\limits_{-\infty}^{\xi+0} \mathsf{P}(H_0; \xi') d\xi' \, B_{w\beta}(\mathbf{R} - \mathbf{R}', \xi - \xi') \left\langle \frac{\delta \varphi\left[\mathbf{u}(\widetilde{\mathbf{R}}, \widetilde{z}, \tau\right]}{\delta u_\beta(\mathbf{R}', \xi', t - 0)} \right\rangle$$

$$- \frac{\partial}{\partial \xi} \int d\mathbf{R}' \int\limits_{-\infty}^{\xi+0} \mathsf{P}(H_0; \xi') d\xi' \, B_{ww}(\mathbf{R} - \mathbf{R}', \xi - \xi') \left\langle \frac{\delta \varphi\left[\mathbf{u}(\widetilde{\mathbf{R}}, \widetilde{z}, \tau\right]}{\delta w(\mathbf{R}', \xi', t - 0)} \right\rangle$$

$$+ \int d\mathbf{R}' \int\limits_{-\infty}^{\xi+0} \mathsf{P}(H_0; \xi') d\xi' \, \frac{\partial B_{\alpha\beta}(\mathbf{R} - \mathbf{R}', \xi - \xi')}{\partial R_\gamma} \frac{\partial}{\partial p_\gamma} p_\alpha \left\langle \frac{\delta \varphi\left[\mathbf{u}(\widetilde{\mathbf{R}}, \widetilde{z}, \tau)\right]}{\delta u_\beta(\mathbf{R}', \xi', t - 0)} \right\rangle$$

$$+ \int d\mathbf{R}' \int\limits_{-\infty}^{\xi+0} \mathsf{P}(H_0; \xi') d\xi' \, \frac{\partial B_{\alpha w}(\mathbf{R} - \mathbf{R}', \xi - \xi')}{\partial R_\gamma} \frac{\partial}{\partial p_\gamma} p_\alpha \left\langle \frac{\delta \varphi\left[\mathbf{u}(\widetilde{\mathbf{R}}, \widetilde{z}, \tau)\right]}{\delta w(\mathbf{R}', \xi', t - 0)} \right\rangle$$

$$+ \int d\mathbf{R}' \int\limits_{-\infty}^{\xi+0} \mathsf{P}(H_0; \xi') d\xi' \, \frac{\partial B_{\alpha\beta}(\mathbf{R} - \mathbf{R}', \xi - \xi')}{\partial \xi} \frac{\partial}{\partial p_\gamma} p_\gamma p_\alpha \left\langle \frac{\delta \varphi\left[\mathbf{u}(\widetilde{\mathbf{R}}, \widetilde{z}, \tau)\right]}{\delta u_\beta(\mathbf{R}', \xi', t - 0)} \right\rangle$$

$$+ \int d\mathbf{R}' \int\limits_{-\infty}^{\xi+0} \mathsf{P}(H_0; \xi') d\xi' \, \frac{\partial B_{\alpha w}(\mathbf{R} - \mathbf{R}', \xi - \xi')}{\partial \xi} \frac{\partial}{\partial p_\gamma} p_\gamma p_\alpha \left\langle \frac{\delta \varphi\left[\mathbf{u}(\widetilde{\mathbf{R}}, \widetilde{z}, \tau\right]}{\delta w(\mathbf{R}', \xi', t - 0)} \right\rangle$$

$$-\int d\mathbf{R}' \int_{-\infty}^{\xi+0} P(H_0; \xi')d\xi' \frac{\partial B_{w\beta}(\mathbf{R}-\mathbf{R}', \xi-\xi')}{\partial R_\beta} \frac{\partial}{\partial p_\beta} \left\langle \frac{\delta\varphi\left[\mathbf{u}(\widetilde{\mathbf{R}}, \widetilde{z}, \tau\right]}{\delta u_\beta(\mathbf{R}', \xi', t-0)} \right\rangle$$

$$-\int d\mathbf{R}' \int_{-\infty}^{\xi+0} P(H_0; \xi')d\xi' \frac{\partial B_{ww}(\mathbf{R}-\mathbf{R}', \xi-\xi')}{\partial R_\beta} \frac{\partial}{\partial p_\beta} \left\langle \frac{\delta\varphi\left[\mathbf{u}(\widetilde{\mathbf{R}}, \widetilde{z}, \tau)\right]}{\delta w(\mathbf{R}', \xi', t-0)} \right\rangle$$

$$-\int d\mathbf{R}' \int_{-\infty}^{\xi+0} P(H_0; \xi')d\xi' \frac{\partial B_{w\beta}(\mathbf{R}-\mathbf{R}', \xi-\xi')}{\partial \xi} \frac{\partial}{\partial p_\beta} p_\beta \left\langle \frac{\delta\varphi\left[\mathbf{u}(\widetilde{\mathbf{R}}, \widetilde{z}, \tau)\right]}{\delta u_\beta(\mathbf{R}', \xi', t-0)} \right\rangle$$

$$-\int d\mathbf{R}' \int_{-\infty}^{\xi+0} P(H_0; \xi')d\xi' \frac{\partial B_{ww}(\mathbf{R}-\mathbf{R}', \xi-\xi')}{\partial \xi} \frac{\partial}{\partial p_\beta} p_\beta \left\langle \frac{\delta\varphi\left[\mathbf{u}(\widetilde{\mathbf{R}}, \widetilde{z}, \tau)\right]}{\delta w(\mathbf{R}', \xi', t-0)} \right\rangle$$

$$(10.20)$$

Provided the uniform initial condition $\xi_0(\mathbf{R}) = 0$ and $\mathbf{p}_0(\mathbf{R}) = 0$, all the single-point statistical characteristics are independent of the spatial coordinate $\mathbf{R}$, i.e.

$$P(\mathbf{R}, t; \xi, \mathbf{p}) \equiv P(t; \xi, \mathbf{p}).$$

From now on, we concern ourselves with the case of uniform initial conditions. Spatial irregularities imply diffusion in the $\{\mathbf{r}\}$ space. However, our focus is on spatial structure formation due to diffusion in the phase space $\{\xi, \mathbf{p}\}$.

Then, rewriting Liouville equation (10.10) in an integral form, one obtains the expressions featuring the corresponding variational derivatives

$$\left\langle \frac{\delta\varphi\left[\mathbf{u}(\widetilde{\mathbf{R}}, \widetilde{z}, \tau)\right]}{\delta u_\beta(\mathbf{R}', \xi', t-0)} \right\rangle_{\mathbf{u}} = \delta(\mathbf{R}-\mathbf{R}') \frac{\partial\delta(\xi-\xi')}{\partial\xi} p_\beta P(t; \xi, \mathbf{p})$$

$$+ \frac{\partial\delta(\mathbf{R}-\mathbf{R}')}{\partial R_\gamma} \delta(\xi-\xi') \frac{\partial}{\partial p_\gamma} p_\beta P(t; \xi, \mathbf{p})$$

$$+ \delta(\mathbf{R}-\mathbf{R}') \frac{\partial\delta(\xi-\xi')}{\partial\xi} \frac{\partial}{\partial p_\gamma} p_\gamma p_\beta P(t; \xi, \mathbf{p}), \quad (10.21)$$

$$\left\langle \frac{\delta\varphi\left[\mathbf{u}(\widetilde{\mathbf{R}}, \widetilde{z}, \tau)\right]}{\delta w(\mathbf{R}', \xi', t-0)} \right\rangle_{\mathbf{u}} = -\frac{\partial}{\partial\xi} \delta(\mathbf{R}-\mathbf{R}') \delta(\xi-\xi') P(t; \xi, \mathbf{p})$$

$$- \frac{\partial\delta(\mathbf{R}-\mathbf{R}')}{\partial R_\gamma} \delta(\xi-\xi') \frac{\partial}{\partial p_\gamma} P(t; \xi, \mathbf{p})$$

$$- \delta(\mathbf{R}-\mathbf{R}') \frac{\partial\delta(\xi-\xi')}{\partial\xi} \frac{\partial}{\partial p_\gamma} p_\gamma P(t; \xi, \mathbf{p}).$$

$$(10.22)$$

One can further simplify equations for the function $P(t; \xi, \mathbf{p})$ by using the scalar parameters $p = |\mathbf{p}|$, $I = |\mathbf{p}|^2$ and going to probability density $P(t; \xi, p)$ or $P(t; \xi, I)$.

Then, incorporating Eqs. (10.21) and (10.22) into Eq. (10.20) and integrating it over spatial coordinates $\mathbf{R}'$, ξ', one can derive the required equation

$$
\begin{aligned}
\frac{\partial P(t; \xi, \mathbf{p})}{\partial t} = {}& -\frac{1}{2} \mathsf{P}(H_0; \xi) \frac{\partial^2 B_{\alpha\beta}(\mathbf{0}, 0)}{\partial R_\alpha \partial R_\beta} \frac{\partial}{\partial p_\gamma} p_\gamma P(t; \xi, \mathbf{p}) \\
& + \mathsf{P}(H_0; \xi) \frac{\partial^2 B_{\alpha w}(\mathbf{0}, 0)}{\partial R_\alpha \partial \xi} \frac{\partial}{\partial p_\gamma} p_\gamma P(t; \xi, \mathbf{p}) \\
& - \frac{1}{2} \mathsf{P}(H_0; \xi) \frac{\partial^2 B_{\alpha\alpha}(\mathbf{0}, 0)}{\partial \xi \partial \xi} \frac{\partial}{\partial p_\gamma} p_\gamma p_\beta^2 P(t; \xi, \mathbf{p}) \\
& + \frac{1}{2} \mathsf{P}(H_0; \xi) \frac{\partial^2 B_{\alpha w}(\mathbf{0}, 0)}{\partial R_\alpha \partial \xi} \frac{\partial}{\partial p_\gamma} p_\gamma P(t; \xi, \mathbf{p}) \\
& + B_{ww}(\mathbf{0}, 0) \frac{\partial^2}{\partial \xi^2} \left[\mathsf{P}(H_0; \xi) P(t; \xi, \mathbf{p}) \right] \\
& + B_{ww}(\mathbf{0}, 0) \frac{\partial}{\partial \xi} \left\{ \frac{\partial \mathsf{P}(H_0; \xi)}{\partial \xi} \frac{\partial}{\partial p_\delta} p_\delta P(t; \xi, \mathbf{p}) \right\} \\
& - \mathsf{P}(H_0; \xi) \frac{\partial^2 B_{\alpha\beta}(\mathbf{0}, 0)}{\partial R_\gamma \partial R_\delta} \frac{\partial}{\partial p_\gamma} p_\alpha \frac{\partial}{\partial p_\delta} p_\beta P(t; \xi, \mathbf{p}) \\
& + \frac{1}{2} \mathsf{P}(H_0; \xi) \frac{\partial B_{\alpha w}(\mathbf{0}, 0)}{\partial R_\alpha \partial \xi} \left[\frac{\partial}{\partial p_\gamma} p_\gamma + \frac{\partial}{\partial p_\gamma} p_\gamma \frac{\partial}{\partial p_\delta} p_\delta \right] P(t; \xi, \mathbf{p}) \\
& - \frac{1}{2} \mathsf{P}(H_0; \xi) \frac{\partial^2 B_{\alpha\alpha}(\mathbf{0}, 0)}{\partial \xi \partial \xi} \frac{\partial}{\partial p_\gamma} p_\gamma \frac{\partial}{\partial p_\delta} p_\delta p_\beta p_\beta P(t; \xi, \mathbf{p}) \\
& + \frac{1}{2} \mathsf{P}(H_0; \xi) \frac{\partial B_{\alpha w}(\mathbf{0}, 0)}{\partial R_\alpha \partial \xi} \frac{\partial}{\partial p_\gamma} p_\gamma p_\delta \frac{\partial}{\partial p_\delta} P(t; \xi, \mathbf{p}) \\
& + \frac{1}{2} \mathsf{P}(H_0; \xi) \frac{\partial^2 B_{w\beta}(\mathbf{0}, 0)}{\partial R_\beta \partial \xi} \left[\frac{\partial}{\partial p_\gamma} p_\gamma + \frac{\partial}{\partial p_\gamma} \frac{\partial}{\partial p_\delta} p_\delta p_\gamma \right] P(t; \xi, \mathbf{p}) \\
& - \frac{1}{2} \mathsf{P}(H_0; \xi) \frac{\partial^2 B_{ww}(\mathbf{0}, 0)}{\partial R_\gamma \partial R_\gamma} \frac{\partial}{\partial p_\delta} \frac{\partial}{\partial p_\delta} P(t; \xi, \mathbf{p}) \\
& + \frac{1}{2} \mathsf{P}(H_0; \xi) \frac{\partial^2 B_{w\beta}(\mathbf{0}, 0)}{\partial \xi \partial R_\beta} \frac{\partial}{\partial p_\gamma} p_\gamma \frac{\partial}{\partial p_\delta} p_\delta P(t; \xi, \mathbf{p}) \\
& - \mathsf{P}(H_0; \xi) \frac{\partial^2 B_{ww}(\mathbf{0}, 0)}{\partial \xi \partial \xi} \left[\frac{\partial}{\partial p_\gamma} p_\gamma + \frac{\partial}{\partial p_\gamma} p_\gamma \frac{\partial}{\partial p_\delta} p_\delta \right] P(t; \xi, \mathbf{p}).
\end{aligned}
$$

$$(10.23)$$

Since we make use of the kinematic approximation, random bottom irregularities appear in the equation in probability density as a monotonic function, $\mathsf{P}(H_0; \xi)$ in the diffusion coefficients. It is clear that the effect of this function will be have maximal in the case of the infinitely deep see, i.e., if $P(\infty; \xi) = 1$. We stress that the influence of the bottom irregularities on forming the hydrodynamic velocity field is not in question.

Given the approximation, we study in the next Section how a complex stochastic velocity field structure affects the statistical properties of the sea surface elevation and its gradient.

10.2.3 Statistical Analysis of the Problem

The result proved to be rather curious: for an infinitely deep sea and a spatially statistically homogeneous problem that corresponds to initial conditions $\xi_0(\mathbf{R})$, $p(\mathbf{R}, 0) = 0$, the situation looks as follows.

1. The sea surface elevation $\xi(\mathbf{R}, t)$ does not correlate with its gradient and obeys the Gaussian distribution

$$P(t; \xi) = \frac{1}{\sqrt{4\pi B_{ww}(\mathbf{0}, 0)t}} \exp\left\{-\frac{\xi^2}{4B_{ww}(\mathbf{0}, 0)t}\right\}, \tag{10.24}$$

which does not depend on the nonlinearity of the input equation (10.6). The variance of the sea surface elevation is given by

$$\sigma_\xi^2(t) = \left\langle \xi^2(\mathbf{R}, t)\right\rangle_\xi = \int_{-\infty}^{\infty} \xi^2 P(t; \xi)d\xi = 2B_{ww}(\mathbf{0}, 0)t.$$

Here, the subscript ξ labels averaging over an ensemble of realizations of field $\xi(\mathbf{R}, t)$. The diffusion coefficient $B_{ww}(\mathbf{0}, 0)$ in expression (10.24) is linked to the correlation function of the vertical velocity $w(\mathbf{R}, z, t)$ by the relationship

$$B_{ww}(\mathbf{0}, 0) = \int_0^{\infty} d\tau\, B_{ww}(\mathbf{0}, 0; \tau),$$

where $B_{ww}(\mathbf{R}, z; t)$ is the correlation function of field $w(\mathbf{R}, z, t)$. The coefficient $B_{ww}(\mathbf{0}, 0)$, in turn, is linked with the variance of random three-dimensional velocity field $\mathbf{u}(\mathbf{r}, t)$ (4.3):

$$B_{ww}(\mathbf{0}, 0) = \frac{1}{3}\int d\mathbf{k}\, \left[2E^s(k) + E^P(k)\right] = \frac{1}{3}\sigma_{\mathbf{u}}^2 \tau_0, \tag{10.25}$$

where τ_0 is the characteristic temporal correlation radius of a random velocity field $\mathbf{u}(\mathbf{r}, t)$.

As a consequence, the complex structure of the velocity field in the lower subspace cannot be the direct cause for stochastic structure formation for the sea surface elevation.

Note that expressions for conditional means follow from formula (10.24):

$$\langle \xi(\mathbf{R}, t) | \xi > 0 \rangle_\xi = \int\limits_0^\infty d\xi \, \xi P(t; \xi) = \sqrt{\frac{1}{\pi} B_{ww}(\mathbf{0}, 0) t}, \tag{10.26}$$

$$\langle \xi(\mathbf{R}, t) | \xi < 0 \rangle_\xi = \int\limits_{-\infty}^0 d\xi \, \xi P(t; \xi) = -\sqrt{\frac{1}{\pi} B_{ww}(\mathbf{0}, 0) t}. \tag{10.27}$$

Here, certainly, $\langle \xi(\mathbf{R}, t) \rangle_\xi = 0$.

2. The probability density $P(t; I)$ the squared modulus of the sea surface elevation gradient $I(\mathbf{R}, t) = \mathbf{p}^2(\mathbf{R}, t)$ is described by an equation of universal form:

$$\frac{\partial}{\partial \tau} P(\tau; I) = \frac{\partial}{\partial I} I(1 + I) P(\tau; I) + 2 \frac{\partial}{\partial I} I \frac{\partial}{\partial I} (1 + I)^2 P(\tau; I), \tag{10.28}$$

where the dimensionless time was introduced:

$$\tau = \frac{2}{15} \left(4 D^{\mathrm{s}} + D^{\mathrm{P}}\right) t, \tag{10.29}$$

and the quantities D^{s} and D^{P} are given by equalities (4.7) with $d = 3$.

We note first that for any fixed point $\widetilde{\mathbf{R}}$ in space, the function $I\left(\tau; \widetilde{\mathbf{R}}\right)$ is a random process in time, whose one-time probability density is independent of $\widetilde{\mathbf{R}}$ and is described by the equation obtained.

Alongside this, in physical space $\{\mathbf{R}\}$ the process of structure formation in the field $I(\tau; \mathbf{R}) = |\mathbf{p}(\tau; \mathbf{R})|^2$ considered as a physical object may take place in the form of closed regions with an augmented gradient concentration, which is clustering described by Eq. (10.28).

A qualitative analysis and estimates presented in Refs. [83, 84] have shown the following. Equation (10.28) is rather complex and comprises two effects. On the one hand, field $I(\tau; \mathbf{R})$ is generated by the random Gaussian velocity field. But, on the other hand, it is parametrically excited by virtue of the dynamics of the input stochastic equations. The random field $I(\tau; \mathbf{R})$ decays with a probability one (i.e., in almost all its realizations) at a sufficiently large time at any point in space and, hence, should undergo clustering in small spatial regions.

The integral probability distribution function $P(\tau; I)$ defined as

$$\Phi(\tau, I) = \int\limits_0^I dI' \, P(\tau; I') = \langle \theta(I - I(\tau; \mathbf{R})) \rangle_\mathbf{u}$$

is the probability $P\left(I\left(\tau;\mathbf{R}\right)<I\right)$ of event $I\left(\tau;\mathbf{R}\right)<I$. The function $\Phi(\tau,I)$ satisfies the equation

$$\frac{\partial}{\partial\tau}\Phi(\tau,I)=I(1+I)\frac{\partial}{\partial I}\Phi(\tau,I)+2I\frac{\partial}{\partial I}(1+I)^2\frac{\partial}{\partial I}\Phi(\tau,I),\qquad(10.30)$$

which follows from Eq. (10.28). Relatedly, the function

$$\widetilde{\Phi(\tau,I)}=\int\limits_{I}^{\infty}dI'\,P(\tau;I')=\langle\theta\left(I\left(\tau;\mathbf{R}\right)-I\right)\rangle_{\mathbf{u}}=1-\Phi(\tau,I)$$

is the probability $P\left(I\left(\tau;\mathbf{R}\right)>I\right)$ of event $I\left(\tau;\mathbf{R}\right)>I$. Because of parametric excitation, function $\Phi(\tau,I)$ rapidly approaches unity with time, while function $\widetilde{\Phi(\tau,I)}$ tends to zero [83, 84].

In general, the area over which the random field $I\left(\tau;\mathbf{R}\right)$ exceeds the fixed level $\overline{I}$, is described by the integral

$$S\left(\tau,\overline{I}\right)=\int d\mathbf{R}\,\theta\left(I\left(\tau;\mathbf{R}\right)-\overline{I}\right),\qquad(10.31)$$

and the total 'mass' of field $I\left(\tau;\mathbf{R}\right)>\overline{I}$ confined within this area is given by

$$I\left(\tau;I>\overline{I}\right)=\int d\mathbf{R}\,I(\tau;\mathbf{R})\,\theta\left(I(\tau;\mathbf{R})-\overline{I}\right).\qquad(10.32)$$

Averaging expressions (10.31) and (10.32) over an ensemble of realizations of random field $I\left(\tau;\mathbf{R}\right)$, we obtain for the mean quantities:

$$\begin{aligned}\langle S\left(\tau,\overline{I}\right)\rangle_I&=\int d\mathbf{R}\,\langle\theta\left(I\left(\tau;\mathbf{R}\right)-\overline{I}\right)\rangle_I,\\\langle I\left(\tau;I>\overline{I}\right)\rangle_I&=\int d\mathbf{R}\,\langle I(\tau;\mathbf{R})\,\theta\left(I(\tau;\mathbf{R})-\overline{I}\right)\rangle_I.\end{aligned}\qquad(10.33)$$

For a spatially homogeneous and isotropic problem, all one-point statistical means are independent of spatial coordinate $\mathbf{R}$, so that the first equality in Eq. (10.33) should be written for the specific quantity

$$s_{\mathrm{hom}}\left(\tau,\overline{I}\right)=\langle\theta\left(I\left(\tau;\mathbf{R}\right)-\overline{I}\right)\rangle_I,\qquad(10.34)$$

which is the probability $P\left(I\left(\tau;\mathbf{R}\right)>\overline{I}\right)$ of event $\left(I\left(\tau;\mathbf{R}\right)>\overline{I}\right.$. According to the ideas of statistical topography of random fields, the function $s_{\mathrm{hom}}\left(\tau,\overline{I}\right)$ for a statistically homogeneous field has a geometric sense of the specific value (i.e., area per unit area, or portion of an area), over which field $I\left(\tau;\mathbf{R}\right)$ exceeds arbitrary given value $\overline{I}$ (see, for example, monographs [28, 29]). However, for a field decaying at almost all points in space, this probability tends to zero, which says that the basic statistical characteristics, such as moment functions, are concentrated on this small portion of an area. The specific square of gradient modulus i concentrated in this region is described by the equality

$$\langle i\,(\tau; I > \overline{I})\rangle_I = \int_{\overline{I}}^{\infty} dI\; I P(\tau; I) = \langle I(t)\rangle_I \left\{ 1 - \frac{1}{\langle I(t)\rangle_I} \int_{0}^{\overline{I}} dI\; I P(\tau; I) \right\}.$$

(10.35)

Hence, it follows that, as time progresses, the expression in Eq. (10.35) to the right of the second equality sign tends to $\langle I(t)\rangle_I$. Accordingly, in this small portion of space, a strong gradient should to all appearance simply *extrude* the free surface both upward (i.e., to form *tall narrow structures*) and downward (i.e., to create *deep narrow troughs—maelstroms* of free surface) as the area where it is confined is decreased. And this should correspond to rare large fluctuations of the Gaussian field of sea surface elevation $\xi(\mathbf{R}, t)$.

10.2.4   Statistical Topography of the Sea Surface Elevation Field

Consider now how the process of stochastic structure formation can be described in terms of the sea surface $\xi(\mathbf{R}, t)$ itself. We know that the values of the random field of squared gradient modulus $I\,(\tau; \mathbf{R})$ exceeding a fixed magnitude $\overline{I}$), i.e., $I\,(\tau; \mathbf{R}) > \overline{I}$, are confined within small spatial regions (10.31) in individual realizations. Namely within this small portion of space, the process of structure formation of the sea surface elevation $\xi(\mathbf{R}, t)$ should take place. Consequently, for the field $\xi(\mathbf{R}, t)$ we should study the probability density in the region of space (10.31), i.e., the conditional probability density of the event $I\,(\tau; \mathbf{R}) > \overline{I}$ under the condition $\xi > 0$. In this case, the fields $\xi(\mathbf{R}, t)$ and $I(\mathbf{R}, t)$ are statistically independent.

   Thus, the integral quantity of interest

$$\int d\mathbf{R}\xi(\mathbf{R}, t)\theta(\xi(\mathbf{R}, t))\theta(I\,(\tau; \mathbf{R}) - \overline{I})$$

averaged over ensembles of realizations of fields $\xi(\mathbf{R}, t)$ and $I\,(\tau; \mathbf{R})$ will define the conditional mean which, for a spatially homogeneous and isotropic problem, according to Eqs. (10.26) and (10.34) takes the form

$$\langle \xi(t)|\xi > 0\rangle\; s_{\text{hom}}\left(\tau, \overline{I}\right) = \sqrt{\frac{1}{\pi} B_{ww}(\mathbf{0}, 0)t}\; s_{\text{hom}}\left(\tau, \overline{I}\right).$$

(10.36)

Taking into account Eq. (10.29), we can express time t in terms of dimensionless time τ and recast conditional mean (10.36) as

$$\langle \xi(\mathbf{R}, t)\theta(\xi(\mathbf{R}, t))\theta(I\,(\tau; \mathbf{R}) - \overline{I})\rangle_{\xi, I} = \sqrt{\frac{5\sigma_{\mathbf{u}}^2 \tau_0}{2\pi\,(4D^s + D^P)}}\, F\,(\tau, \overline{I})\,, \quad (10.37)$$

where the dimensionless function

$$F\,(\tau; \overline{I}) = \sqrt{\tau}\, s_{\text{hom}}\,(\tau, \overline{I})\,, \quad (10.38)$$

and the coefficients D^s and D^P given by Eqs. (4.7) are expressed through statistical parameters of velocity fields using Eq. (4.8).

We note that in incompressible hydrodynamic turbulence ($D^P = 0$), Eq. (10.37) is substantially simplified and, with account for equalities (4.8), acquires the universal form

$$\langle \xi(\mathbf{R}, t)\theta(\xi(\mathbf{R}, t))\theta(I\,(\tau; \mathbf{R}) - \overline{I})\rangle_{\xi, I} = \sqrt{\frac{5}{4\pi}}\, l_\omega F\,(\tau, \overline{I})\,, \quad (10.39)$$

where $l_\omega = \sigma_{\mathbf{u}}/\sigma_\omega$ is the spatial correlation radius of the vorticity field.

For small times, this conditional mean grows with time, but then decays because of a rapid decrease in function $s_{\text{hom}}\,(\tau, \overline{I})$ passing through the maximum which characterizes the mean maximum amplitude of anomalous structure formation on the sea surface.

In a similar way, one may compute other conditional moment functions like

$$\langle \xi^n(\mathbf{R}, t) | \xi > 0;\, I\,(\tau; \mathbf{R}) > \overline{I}\rangle\,.$$

It is also possible to consider the integral quantity

$$\int d\mathbf{R}\, \xi(\mathbf{R}, t)\theta(-\xi(\mathbf{R}, t))\theta(I\,(\tau; \mathbf{R}) - \overline{I}),$$

which, on averaging over ensembles of realizations of fields $\xi(\mathbf{R}, t)$ and $I\,(\tau; \mathbf{R})$, defines the conditional mean for the spatially homogeneous and isotropic case:

$$\langle \xi(t) | \xi < 0\rangle\, s_{\text{hom}}\,(\tau, \overline{I}) = -\sqrt{\frac{1}{\pi} B_{ww}(\mathbf{0}, 0)t}\, s_{\text{hom}}\,(\tau, \overline{I})\,. \quad (10.40)$$

For small times, conditional mean (10.40) decays with time, but will increase further because of a rapid decrease in positive function $s_{\text{hom}}\,(\tau, \overline{I})$, passing through a minimum value which characterizes the mean minimum amplitude of anomalous negative structure formation on the sea surface (i.e., deep trough).

Thus, based on the equations obtained in Refs. [83, 84], it is shown that in the process of structure formation of the gradient modulus, the field of sea surface elevation increases in small regions which comprise the entire gradient, reaching a maximum and then beginning to decrease. Thus, the structures studied may reach a substantial amplitude on the surface of a fluid, but are limited both in magnitude and lifetime.

Let us make a remark concerning the interpretation of statistical means from the standpoint of statistical topography. Formulas (10.36) and (10.40) show that sea surface elevations differing from zero are confined in the regions where the modulus of the gradient undergoes clustering, i.e., in such regions where this modulus exceeds an arbitrarily small value of $\overline{I}$. Note that at $\overline{I} = 0$ Eqs. (10.36) and (10.40) lose their meaning since they cease to describe the distribution of the elevation in space and become, therefore, expressions for ordinary means, the interpretation of which from the standpoint of behavior in individual realizations does not convey useful information.

We also note that the estimates obtained in Refs. [83, 84] and general features of the sea surface elevation amplitude only point to the existence of an effect. More detailed characteristics and additional verification can be furnished by numerically solving Eqs. (10.28) and (10.30), and also by numerical simulations of the input equation (10.6).

Conclusion

We have considered the processes of stochastic structure formation in the two-dimensional geophysical fluid dynamics based on statistical analysis of Gaussian random fields, as well as stochastic structure formation in dynamic systems with parametric excitation of positive random fields $f(\mathbf{r}, t)$ described by partial differential equations. We also considered two examples of stochastic structure formation in dynamic systems with parametric excitation in the presence of the Gaussian pumping. Such structure formation in dynamic systems with parametric excitation in space and time either happens or not! However, if it occurs in space, then this almost always happens (exponentially fast) in individual realizations, i.e., with a unit probability, and for the spatially homogeneous statistical case consists in the following:

(**1**) the field decays at almost all points in space with time (clearly, with fluctuations superimposed);

(**2**) the small regions where this field is concentrated (clustered) develop in space, and stochastic structure formation is caused by diffusion of random field $f(\mathbf{r}, t)$ in its phase space $\{f\}$.

In the case considered, clustering of the field $f(\mathbf{r}, t)$ of any nature is a general feature of dynamic fields, and one may claim that structure formation is the *Law of Nature* for arbitrary random fields of such type.

In this study, we clarified conditions under which such structure formation takes place. It is worth noting that these conditions have a transparent physical-mathematical sense and are described at a sufficiently elementary mathematical level by resorting to the ideas of statistical topography.

The author is indebted to K.V. Koshel' for his longstanding help with numerical modeling of dynamic systems and numerical solution of various equations, and also for joint analytical work on statistical analysis of the problem involving stochastic structures on the sea surface.

This study was supported by the Russian Science Foundation, Project No. 14-27-00134.

Appendix
Elements of Mathematical Tools for Describing Coherent Phenomena

A. Solution Dependence on Problem Type, Medium Parameters, and Initial Data

Earlier, in Sect. 3 we considered a number of dynamic systems described by both ordinary and partial differential equations. Many applications concerning research of statistical characteristics of the solutions to these equations require the knowledge of the solution dependence (generally, in the functional form) on the medium parameters appeared in the equations as coefficients and initial values. Some properties appear common of all such dependencies, and two of them are of special interest in the context of statistical descriptions. We illustrate these dependencies by the example of the simplest problem, namely, the system of ordinary differential equations (3.1) that describes particle dynamics under random velocity field and which we reformulate in the form of the nonlinear integral equation

$$\mathbf{r}(t) = \mathbf{r}_0 + \int_{t_0}^{t} d\tau \mathbf{U}(\mathbf{r}(\tau), \tau). \tag{A.1}$$

The solution to Eq. (A.1) functionally depends on vector field $\mathbf{U}(\mathbf{r}', \tau)$ and initial values $\mathbf{r}_0$, t_0.

A.1 Functional Representation of Problem Solution

A.1.1 Variational (Functional) Derivatives

Recall first the general definition of a functional. One says that a functional is given if a rule is fixed that associates a number to every function from certain function family. Below, we give some examples of functionals:

© Springer International Publishing AG 2017
V.I. Klyatskin, *Fundamentals of Stochastic Nature Sciences*,
Understanding Complex Systems, DOI 10.1007/978-3-319-56922-2

$$\textbf{(a)} \quad F[\varphi(\tau)] = \int\limits_{t_1}^{t_2} d\tau a(\tau)\varphi(\tau),$$

where $a(t)$ is the given (fixed) function and limits t_1 and t_2 can be both finite and infinite. This is the linear functional.

$$\textbf{(b)} \quad F[\varphi(\tau)] = \int\limits_{t_1}^{t_2} \int\limits_{t_1}^{t_2} d\tau_1 d\tau_2 B(\tau_1, \tau_2)\varphi(\tau_1)\varphi(\tau_2),$$

where $B(t_1, t_2)$ is the given (fixed) function. This is the quadratic functional.

$$\textbf{(c)} \quad F[\varphi(\tau)] = f\left(\Phi[\varphi(\tau)]\right),$$

where $f(x)$ is the given function and quantity $\Phi[\varphi(\tau)]$ is the functional.

Estimate the difference between the values of a functional calculated for functions $\varphi(\tau)$ and $\varphi(\tau) + \delta\varphi(\tau)$ for $t - \dfrac{\Delta\tau}{2} < \tau < t + \dfrac{\Delta\tau}{2}$ (see Fig. A.1).

The variation of a functional is defined as the linear (in $\delta\varphi(\tau)$) portion of the difference

$$\delta F[\varphi(\tau)] = \{F\left[\varphi(\tau) + \delta\varphi(\tau)\right] - F[\varphi(\tau)]\}.$$

The limit

$$\frac{\delta F[\varphi(\tau)]}{\delta\varphi(t)dt} = \lim_{\Delta\tau \to 0} \frac{\delta F[\varphi(\tau)]}{\int\limits_{\Delta\tau} d\tau \delta\varphi(\tau)} \tag{A.2}$$

is called the *variational* (or *functional*) *derivative* (see, e.g., [41]).

Fig. A.1 To definition of variational derivative

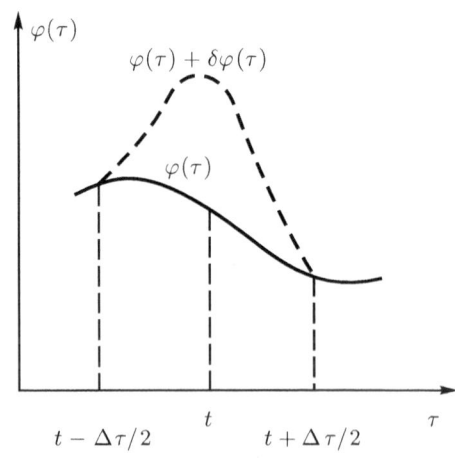

For short, we will use notation $\dfrac{\delta F[\varphi(\tau)]}{\delta\varphi(t)}$ instead of $\dfrac{\delta F[\varphi(\tau)]}{\delta\varphi(t)dt}$.

Note that, if we use function $\delta\varphi(\tau) = \alpha\delta(\tau)$, where $\delta(\tau)$ is the Dirac delta function, then Eq. (A.2) can be represented in the form of the ordinary derivative

$$\frac{\delta F[\varphi(\tau)]}{\delta\varphi(t)} = \lim_{\alpha\to 0}\frac{d}{d\alpha}F[\varphi(\tau) + \alpha\delta(\tau - t)].$$

The variational derivative of functional $F[\varphi(\tau)]$ is again the functional of $\varphi(\tau)$, which depends additionally on point t as a parameter. As a result, this variational derivative will have two types of derivatives; one can differentiate it in the ordinary sense with respect to parameter t and in the functional sense with respect to $\varphi(\tau)$ at point $\tau = t'$ thus obtaining the second variational derivative of the initial functional

$$\frac{\delta^2 F[\varphi(\tau)]}{\delta\varphi(t')\delta\varphi(t)} = \frac{\delta}{\delta\varphi(t')}\left[\frac{\delta F[\varphi(\tau)]}{\delta\varphi(t)}\right].$$

The second variational derivative will now be the functional of $\varphi(\tau)$ dependent on two points t and t', and so forth.

Determine the variational derivatives of functionals (a), (b), and (c).

In the case (a), we have

$$\delta F[\varphi(\tau)] = F[\varphi(\tau) + \delta\varphi(\tau)] - F[\varphi(\tau)] = \int_{t-\frac{\Delta\tau}{2}}^{t+\frac{\Delta\tau}{2}} d\tau a(\tau)\delta\varphi(\tau).$$

If function $a(t)$ is continuous on segment $\Delta\tau$, then, by the average theorem,

$$\delta F[\varphi(\tau)] = a(t')\int_{\Delta\tau} d\tau\delta\varphi(\tau),$$

where point t' belongs to segment $\left[t - \dfrac{\Delta\tau}{2}, t + \dfrac{\Delta\tau}{2}\right]$. Consequently,

$$\frac{\delta F[\varphi(\tau)]}{\delta\varphi(t)} = \lim_{\Delta\tau\to 0} a(t') = a(t). \tag{A.3}$$

In the case (b), we obtain similarly

$$\frac{\delta F[\varphi(\tau)]}{\delta\varphi(t)} = \int_{t_1}^{t_2} d\tau\, [B(\tau,t) + B(t,\tau)]\,\varphi(\tau)\quad (t_1 < t < t_2).$$

Note that function $B(\tau_1, \tau_2)$ can always be assumed a symmetric function of its arguments here.

In the case (**c**), we have

$$F[\varphi(\tau) + \delta\varphi(\tau)] = f\left(\Phi[\varphi(\tau)]\right) + \frac{\partial f(\Phi[\varphi(\tau)])}{\partial\Phi}\delta\Phi[\varphi(\tau)] + \cdots$$

$$= F[\varphi(\tau)] + \frac{\partial f(\Phi[\varphi(\tau)])}{\partial\Phi}\delta\Phi[\varphi(\tau)] + \cdots$$

and, consequently,

$$\frac{\delta}{\delta\varphi(t)} f\left(\Phi[\varphi(\tau)]\right) = \frac{\partial f(\Phi[\varphi(\tau)])}{\partial\Phi}\frac{\delta}{\delta\varphi(t)}\Phi[\varphi(\tau)]. \tag{A.4}$$

Consider now functional $\Phi[\varphi(\tau)] = F_1[\varphi(\tau)]F_2[\varphi(\tau)]$. We have

$$\delta\Phi[\varphi(\tau)] = F_1[\varphi(\tau) + \delta\varphi(\tau)]F_2[\varphi(\tau) + \delta\varphi(\tau)] - F_1[\varphi(\tau)]F_2[\varphi(\tau)]$$

$$= F_1[\varphi(\tau)]\delta F_2[\varphi(\tau)] + F_2[\varphi(\tau)]\delta F_1[\varphi(\tau)]$$

and, consequently,

$$\frac{\delta}{\delta\varphi(t)}F_1[\varphi(\tau)]F_2[\varphi(\tau)] = F_1[\varphi(\tau)]\frac{\delta}{\delta\varphi(t)}F_2[\varphi(\tau)] + F_2[\varphi(\tau)]\frac{\delta}{\delta\varphi(t)}F_1[\varphi(\tau)]. \tag{A.5}$$

We can define the expression for the variational derivative of functional $\varphi(\tau_0)$ with respect to function $\varphi(t)$ by the formal relationship

$$\frac{\delta\varphi(\tau_0)}{\delta\varphi(t)} = \delta(\tau_0 - t). \tag{A.6}$$

Formula (A.6) can be proved, for example, by considering the linear functional of the form

$$F[\varphi(\tau)] = \frac{1}{\sqrt{2\pi}\sigma}\int\limits_{-\infty}^{\infty} d\tau\varphi(\tau)\exp\left\{-\frac{(\tau - \tau_0)^2}{2\sigma^2}\right\}. \tag{A.7}$$

According to Eq. (A.3), the variational derivative of this functional has the form

$$\frac{\delta}{\delta\varphi(t)}F[\varphi(\tau)] = \frac{1}{\sqrt{2\pi}\sigma}\exp\left\{-\frac{(t - \tau_0)^2}{2\sigma^2}\right\}. \tag{A.8}$$

Performing now formal limit process $\sigma \to 0$ in Eqs. (A.7) and (A.8), we obtain the desired formula (A.6). Moreover,

$$\frac{\delta F[\varphi(\tau)]}{\delta\varphi(t)} = \frac{\partial F[\varphi(\tau)]}{\partial\varphi(\tau)}\frac{\delta\varphi(\tau)}{\delta\varphi(t)} = \frac{\partial F[\varphi(\tau)]}{\partial\varphi(\tau)}\delta(\tau - t).$$

Formula (A.6) is very convenient for functional differentiation of functionals explicitly dependent on $\varphi(\tau)$. Indeed, for the quadratic functional (**b**), we have

$$\frac{\delta}{\delta\varphi(t)} \int_{t_1}^{t_2} \int_{t_1}^{t_2} d\tau_1 d\tau_2 B(\tau_1, \tau_2) \varphi(\tau_1)\varphi(\tau_2)$$

$$\overset{(250)}{=} \int_{t_1}^{t_2} \int_{t_1}^{t_2} d\tau_1 d\tau_2 B(\tau_1, \tau_2) \left[\frac{\delta\varphi(\tau_1)}{\delta\varphi(t)}\varphi(\tau_2) + \varphi(\tau_1)\frac{\delta\varphi(\tau_2)}{\delta\varphi(t)} \right]$$

$$\overset{(251)}{=} \int_{t_1}^{t_2} d\tau \, [B(t, \tau) + B(\tau, t)]\,\varphi(\tau) \quad (t_1 < t < t_2).$$

Consider the functional

$$F[\varphi(\tau)] = \int_{t_1}^{t_2} d\tau L\left(\tau, \varphi(\tau), \frac{d\varphi(\tau)}{d\tau}\right)$$

as another example. In this case,

$$\frac{\delta}{\delta\varphi(t)}F[\varphi(\tau)] \overset{(249)}{=} \int_{t_1}^{t_2} d\tau \left[\frac{\partial L\left(\tau, \varphi(\tau), \frac{d\varphi(\tau)}{d\tau}\right)}{\partial\varphi(\tau)} + \frac{\partial L\left(\tau, \varphi(\tau), \frac{d\varphi(\tau)}{d\tau}\right)}{\partial\dot\varphi(\tau)}\frac{d}{d\tau} \right] \frac{\delta\varphi(\tau)}{\delta\varphi(t)}$$

$$\overset{(251)}{=} \int_{t_1}^{t_2} d\tau \left[\frac{\partial L\left(\tau, \varphi(\tau), \frac{d\varphi(\tau)}{d\tau}\right)}{\partial\varphi(\tau)} + \frac{\partial L\left(\tau, \varphi(\tau), \frac{d\varphi(\tau)}{d\tau}\right)}{\partial\dot\varphi(\tau)}\frac{d}{d\tau} \right] \delta(\tau - t)$$

$$= \left(-\frac{d}{dt}\frac{\partial}{\partial\dot\varphi(t)} + \frac{\partial}{\partial\varphi(t)}\right) L\left(t, \varphi(t), \frac{d\varphi(t)}{dt}\right),$$

where $\dot\varphi(t) = \frac{d}{dt}\varphi(t)$ if point t belongs to interval (t_1, t_2).

Just as a function can be expanded in the Taylor series, a functional $F[\varphi(\tau) + \eta(\tau)]$ can be expanded in the functional Taylor series in function $\eta(\tau)$

$$F[\varphi(\tau) + \eta(\tau)] = F[\varphi(\tau)] + \int_{-\infty}^{\infty} dt \frac{\delta F[\varphi(\tau)]}{\delta\varphi(t)}\eta(t)$$

$$+ \frac{1}{2!} \int_{-\infty}^{\infty} \int_{-\infty}^{\infty} dt_1 dt_2 \frac{\delta^2 F[\varphi(\tau)]}{\delta\varphi(t_1)\delta\varphi(t_2)}\eta(t_1)\eta(t_2) + \cdots . \quad (A.9)$$

Note that the operator expression

$$1 + \int\limits_{-\infty}^{\infty} dt\eta(t)\frac{\delta}{\delta\varphi(t)} + \frac{1}{2!}\int\limits_{-\infty}^{\infty}\int\limits_{-\infty}^{\infty} dt_1 dt_2\eta(t_1)\eta(t_2)\frac{\delta^2}{\delta\varphi(t_1)\delta\varphi(t_2)} + \cdots$$

$$= 1 + \int\limits_{-\infty}^{\infty} dt\eta(t)\frac{\delta}{\delta\varphi(t)} + \frac{1}{2!}\left[\int\limits_{-\infty}^{\infty} dt\eta(t)\frac{\delta}{\delta\varphi(t)}\right]^2 + \cdots \quad (A.10)$$

can be written shortly as the operator

$$\exp\left\{\int\limits_{-\infty}^{\infty} dt\eta(t)\frac{\delta}{\delta\varphi(t)}\right\}, \quad (A.11)$$

whose action should be treated precisely in the sense of expansion (A.10). Using this operator, we can rewrite Eq. (A.9) in the form

$$F[\varphi(\tau) + \eta(\tau)] = e^{\int\limits_{-\infty}^{\infty} dt\eta(t)\frac{\delta}{\delta\varphi(t)}} F[\varphi(\tau)], \quad (A.12)$$

which enables us to interpret operator (A.11) as the functional shift operator.

Consider now functional $F[t; \varphi(\tau)]$ dependent on parameter t. We can differentiate this functional with respect to t and determine its variational derivative with respect to $\varphi(t')$, as well. One can easily see that these operations commute, i.e., the equality

$$\frac{\partial}{\partial t}\frac{\delta F[t; \varphi(\tau)]}{\delta\varphi(t')} = \frac{\delta}{\delta\varphi(t')}\frac{\partial F[t; \varphi(\tau)]}{\partial t} \quad (A.13)$$

holds. If the domain of τ is independent of t, the validity of Eq. (A.13) is obvious. Otherwise, for example, for functionals $F[t; \varphi(\tau)]$ with $0 \le \tau \le t$, the validity of Eq. (A.13) can be checked on by expanding functional $F[t; \varphi(\tau)]$ in the functional Taylor series.

A.1.2 Principle of Dynamic Causality

Vary Eq. (A.1) with respect to field $\mathbf{U}(\mathbf{r}, t)$. Assuming that the initial position $\mathbf{r}_0$ is independent of field $\mathbf{U}$, we obtain the equation linear in variational derivative (the linear variational differential equation)

$$\frac{\delta r_i(t)}{\delta U_j(\mathbf{r}, t')} = \delta_{ij}\delta(\mathbf{r} - \mathbf{r}(t'))\theta(t' - t_0)\theta(t - t') + \int\limits_{t_0}^{t} d\tau\frac{\partial U_i(\mathbf{r}(\tau), \tau)}{\partial r_k}\frac{\delta r_k(\tau)}{\delta U_j(\mathbf{r}, t')},$$

$$(A.14)$$

where $\delta\left(\mathbf{r}-\mathbf{r}'\right)$ is the Dirac delta function, and $\theta\left(z\right)$ is the Heaviside step function. From Eq. (A.14) follows that

$$\frac{\delta r_i\left(t\right)}{\delta U_j\left(\mathbf{r},t'\right)}=0 \quad \text{for } t' > t \text{ or } t' < t_0,$$ (A.15)

which means that solution $\mathbf{r}(t)$ to the dynamic problem (A.1) as a functional of field $\mathbf{U}\left(\mathbf{r},t'\right)$ depends on $\mathbf{U}\left(\mathbf{r},t'\right)$ only for $t_0 < t' < t$. Consequently, function $\mathbf{r}(t)$ will remain unchanged if field $\mathbf{U}\left(\mathbf{r},t'\right)$ varies outside the interval (t_0, t), i.e., for $t' < t_0$ or $t' > t$. We will call condition (A.15) the *dynamic causality condition*. Taking this condition into account, we can rewrite Eq. (A.14) in the form

$$\frac{\delta r_i\left(t\right)}{\delta U_j\left(\mathbf{r},t'\right)}=\delta_{ij}\delta\left(\mathbf{r}-\mathbf{r}\left(t'\right)\right)\theta\left(t'-t_0\right)\theta\left(t-t'\right)+\int\limits_{t'}^{t}d\tau\frac{\partial U_i\left(\mathbf{r}\left(\tau\right),\tau\right)}{\partial r_k}\frac{\delta r_k\left(\tau\right)}{\delta U_j\left(\mathbf{r},t'\right)}.$$

(A.16)

As a consequence, limit $t \to t' + 0$ yields the equality

$$\left.\frac{\delta r_i\left(t\right)}{\delta U_j\left(\mathbf{r},t'\right)}\right|_{t=t'+0}=\delta_{ij}\delta\left(\mathbf{r}-\mathbf{r}\left(t'\right)\right).$$ (A.17)

Integral equation (A.16) in variational derivative is obviously equivalent to the linear differential equation with the initial value

$$\frac{\partial}{\partial t}\left(\frac{\delta r_i\left(t\right)}{\delta U_j\left(\mathbf{r},t'\right)}\right)=\frac{\partial U_i\left(\mathbf{r}\left(t\right),t\right)}{\partial r_k}\left(\frac{\delta r_k\left(t\right)}{\delta U_j\left(\mathbf{r},t'\right)}\right),$$

$$\left.\frac{\delta r_i\left(t\right)}{\delta U_j\left(\mathbf{r},t'\right)}\right|_{t=t'}=\delta_{ij}\delta\left(\mathbf{r}-\mathbf{r}\left(t'\right)\right).$$

(A.18)

The dynamic causality condition is the general property of problems described by differential equations with initial values. The boundary-value problems possess no such property. Indeed, in the case of problem (3.12), (3.13) that describes propagation of a plane wave in a layer of inhomogeneous medium, wavefield $u(x)$ at point x and reflection and transmission coefficients depend functionally on function $\varepsilon(x)$ for all x of layer (L_0, L). However, using the imbedding method, we can convert this problem into the initial-value problem with respect to an auxiliary parameter L and make use the causality property in terms of the equations of the imbedding method.

A.2 Solution Dependence on Problem's Parameters

A.2.1 Solution Dependence on Initial Data

Here, we will use the vertical bar symbol to isolate the dependence of solution $\mathbf{r}(t)$ to Eq. (A.1) on the initial parameters $\mathbf{r}_0$ and t_0:

$$\mathbf{r}(t) = \mathbf{r}(t|\mathbf{r}_0, t_0), \quad \mathbf{r}_0 = \mathbf{r}(t_0|\mathbf{r}_0, t_0).$$

Let us differentiate Eq. (A.1) with respect to parameters r_{0k} and t_0. As a result, we obtain linear equations for Jacobi's matrix $\dfrac{\partial}{\partial r_{0k}} r_i(t|\mathbf{r}_0, t_0)$ and quantity $\dfrac{\partial}{\partial t_0} r_i(t|\mathbf{r}_0, t_0)$

$$\frac{\partial r_i(t|\mathbf{r}_0, t_0)}{\partial r_{0k}} = \delta_{ik} + \int_{t_0}^t d\tau \frac{\partial U_i(\mathbf{r}(\tau), \tau)}{\partial r_j} \frac{\partial r_j(\tau|\mathbf{r}_0, t_0)}{\partial r_{0k}},$$

$$\frac{\partial r_i(t|\mathbf{r}_0, t_0)}{\partial t_0} = -U_i(\mathbf{r}_0(t_0), t_0) + \int_{t_0}^t d\tau \frac{\partial U_i(\mathbf{r}(\tau), \tau)}{\partial r_j} \frac{\partial r_j(\tau|\mathbf{r}_0, t_0)}{\partial t_0}.$$

$$(A.19)$$

Multiplying now the first of these equations by $U_k(\mathbf{r}_0(t), t)$, summing over index k, adding the result to the second equation, and introducing the vector function

$$F_i(t|\mathbf{r}_0, t_0) = \left(\frac{\partial}{\partial t_0} + \mathbf{U}(\mathbf{r}_0, t_0) \frac{\partial}{\partial \mathbf{r}_0} \right) r_i(t|\mathbf{r}_0, t_0),$$

we obtain that this function satisfies the linear homogeneous equation

$$F_i(t|\mathbf{r}_0, t_0) = \int_{t_0}^t d\tau \frac{\partial U_i(\mathbf{r}(\tau), \tau)}{\partial r_k} F_k(\tau|\mathbf{r}_0, t_0). \qquad (A.20)$$

Differentiating this equation with respect to time, we arrive at the ordinary differential equation

$$\frac{\partial}{\partial t} F_i(t|\mathbf{r}_0, t_0) = \frac{\partial U_i(\mathbf{r}(t), t)}{\partial r_k} F_k(t|\mathbf{r}_0, t_0)$$

with the initial condition $F_i(t_0|\mathbf{r}_0, t_0) = 0$ at $t = t_0$, which follows from Eq. (A.20); as a consequence, we have $F_i(t|\mathbf{r}_0, t_0) \equiv 0$. Therefore, we obtain the equality

$$\left(\frac{\partial}{\partial t_0} + \mathbf{U}(\mathbf{r}_0, t_0) \frac{\partial}{\partial \mathbf{r}_0} \right) r_i(t|\mathbf{r}_0, t_0) = 0, \qquad (A.21)$$

which can be considered as the linear partial differential equation with the derivatives with respect to variables $\mathbf{r}_0$, t_0 and the initial value at $t_0 = t$

$$\mathbf{r}(t|\mathbf{r}_0, t) = \mathbf{r}_0. \tag{A.22}$$

The variable t appears now in problem (A.21), (A.22) as a parameter.

Equation (A.21) is solved using the time direction inverse to that used in solving problem (3.1); for this reason, we will call it the *backward equation*.

Equation (A.21) with the initial condition (A.22) can be rewritten as the integral equation

$$\mathbf{r}(t|\mathbf{r}_0, t_0) = \mathbf{r}_0 + \int_{t_0}^{t} d\tau \left(\mathbf{U}(\mathbf{r}_0, \tau) \frac{\partial}{\partial \mathbf{r}_0} \right) \mathbf{r}(t|\mathbf{r}_0, \tau). \tag{A.23}$$

Varying now Eq. (A.23) with respect to function $U_j(\mathbf{r}', t')$, we obtain the integral equation

$$\frac{\delta r_i(t|\mathbf{r}_0, t_0)}{\delta U_j(\mathbf{r}', t')} = \delta(\mathbf{r}_0 - \mathbf{r}')\theta(t' - t_0)\theta(t - t') \frac{\partial r_i(t|\mathbf{r}_0, t')}{\partial r_{j0}}$$

$$+ \int_{t_0}^{t} d\tau \left(\mathbf{U}(\mathbf{r}_0, \tau) \frac{\partial}{\partial \mathbf{r}_0} \right) \frac{\delta r_i(t|\mathbf{r}_0, \tau)}{\delta U_j(\mathbf{r}', t')}, \tag{A.24}$$

from which follows that

$$\frac{\delta r_i(t|\mathbf{r}_0, t_0)}{\delta U_j(\mathbf{r}', t')} = 0, \quad \text{if } t' > t \text{ or } t' < t_0,$$

which means that function $\mathbf{r}(t|\mathbf{r}_0, t_0)$ also possesses the property of dynamic causality with respect to parameter t_0 and Eq. (A.24) can be rewritten in the form (for $t_0 < t' < t$)

$$\frac{\delta r_i(t|\mathbf{r}_0, t_0)}{\delta U_j(\mathbf{r}', t')} = \delta(\mathbf{r}_0 - \mathbf{r}') \frac{\partial r_i(t|\mathbf{r}_0, t')}{\partial r_{j0}} + \int_{t_0}^{t'} d\tau \left(\mathbf{U}(\mathbf{r}_0, \tau) \frac{\partial}{\partial \mathbf{r}_0} \right) \frac{\delta r_i(t|\mathbf{r}_0, \tau)}{\delta U_j(\mathbf{r}', t')}. \tag{A.25}$$

Setting now $t' \to t_0 + 0$, we obtain the equality

$$\left. \frac{\delta r_i(t|\mathbf{r}_0, t_0)}{\delta U_j(\mathbf{r}, t')} \right|_{t'=t_0+0} = \delta(\mathbf{r}_0 - \mathbf{r}) \frac{\partial r_i(t|\mathbf{r}_0, t_0)}{\partial r_{0j}}. \tag{A.26}$$

A.2.2 Imbedding Method for Boundary-Value Problems

Consider first boundary-value problems formulated in terms of ordinary differential equations. The *imbedding method* (or *invariant imbedding method*, as it is usually called in mathematical literature) offers a possibility of reducing boundary-value problems at hand to the evolution-type initial-value problems possessing the property of dynamic causality with respect to an auxiliary parameter.

The idea of this method was first suggested by V.A. Ambartsumyan (the so-called *Ambartsumyan invariance principle*) for solving the equations of linear theory of radiative transfer. Further, mathematicians grasped this idea and used it to convert boundary-value (nonlinear, in the general case) problems into evolution-type initial-value problems that are more convenient for simulations. Several monographs (see, e.g., [28, 29]) deal with this method and consider both physical and computational aspects.

Consider the dynamic system described in terms of the system of ordinary differential equations

$$\frac{d}{dt}\mathbf{x}(t) = \mathbf{F}\left(t, \mathbf{x}(t)\right), \tag{A.27}$$

defined on segment $t \in [0, T]$ with the boundary conditions

$$g\mathbf{x}(0) + h\mathbf{x}(T) = \mathbf{v}, \tag{A.28}$$

where g and h are the constant matrixes.

Dynamic problem (A.27), (A.28) possesses no dynamic causality property, which means that the solution $\mathbf{x}(t)$ to this problem at instant t functionally depends on external forces $\mathbf{F}\left(\tau, \mathbf{x}(\tau)\right)$ for all $0 \leq \tau \leq T$. Moreover, even boundary values $\mathbf{x}(0)$ and $\mathbf{x}(T)$ are functionals of field $\mathbf{F}\left(\tau, \mathbf{x}(\tau)\right)$. The absence of dynamic causality in problem (A.27), (A.28) prevents us from using the known statistical methods of analyzing statistical characteristics of the solution to Eq. (A.27) if external force functional $\mathbf{F}\left(t, \mathbf{x}\right)$ is the random space- and time-domain field. Introducing the one-time probability density $P(t; \mathbf{x})$ of the solution to Eq. (A.27), we can easily see that condition (A.28) is insufficient for determining the value of this probability at any point. The boundary condition imposes only certain functional restriction.

Note that the solution to problem (A.27), (A.28) parametrically depends on T and $\mathbf{v}$, i.e., $\mathbf{x}(t) = \mathbf{x}(t; T, \mathbf{v})$. We introduce functions

$$\mathbf{R}(T, \mathbf{v}) = \mathbf{x}(T; T, \mathbf{v}), \quad \mathbf{S}(T, \mathbf{v}) = \mathbf{x}(0; T, \mathbf{v})$$

that describe the boundary values of the solution to Eq. (A.27).

Differentiate Eq. (A.27) with respect to T and $\mathbf{v}$. We obtain two linear equations in the corresponding derivatives

$$\frac{d}{dt}\frac{\partial x_i(t; T, \mathbf{v})}{\partial T} = \frac{\partial F_i(t, \mathbf{x})}{\partial x_l}\frac{\partial x_l(t; T, \mathbf{v})}{\partial T},$$

$$\frac{d}{dt}\frac{\partial x_i(t; T, \mathbf{v})}{\partial v_k} = \frac{\partial F_i(t, \mathbf{x})}{\partial x_l}\frac{\partial x_l(t; T, \mathbf{v})}{\partial v_k}.$$

(A.29)

These equations are identical in form; consequently, we can expect that their solutions are related by the linear expression

$$\frac{\partial x_i(t; T, \mathbf{v})}{\partial T} = \lambda_k(T, \mathbf{v})\frac{\partial x_i(t; T, \mathbf{v})}{\partial v_k}$$

(A.30)

if vector quantity $\lambda(T, \mathbf{v})$ is such that boundary conditions (A.28) are satisfied and the solution is unique. To determine vector quantity $\lambda(T, \mathbf{v})$, we first set $t = 0$ in Eq. (A.30) and multiply the result by matrix g; then, we set $t = T$ and multiply the result by matrix h; and, finally, we combine the obtained expressions. Taking into account Eq. (A.28), we obtain

$$g\frac{\partial \mathbf{x}(0; T, \mathbf{v})}{\partial T} + h\left.\frac{\partial \mathbf{x}(t; T, \mathbf{v})}{\partial T}\right|_{t=T} = \lambda(T, \mathbf{v}).$$

In view of the fact that

$$\left.\frac{\partial \mathbf{x}(t; T, \mathbf{v})}{\partial T}\right|_{t=T} = \frac{\partial \mathbf{x}(T; T, \mathbf{v})}{\partial T} - \left.\frac{\partial \mathbf{x}(t; T, \mathbf{v})}{\partial t}\right|_{t=T} = \frac{\partial \mathbf{R}(T, \mathbf{v})}{\partial T} - \mathbf{F}(T, \mathbf{R}(T, \mathbf{v}))$$

(with allowance for Eq. (A.27)), we obtain the desired expression for quantity $\lambda(T, \mathbf{v})$,

$$\lambda(T, \mathbf{v}) = -h\mathbf{F}(T, \mathbf{R}(T, \mathbf{v})).$$

(A.31)

Expression (A.30) with parameter $\lambda(T, \mathbf{v})$ defined by Eq. (A.31), i.e., the expression

$$\frac{\partial x_i(t; T, \mathbf{v})}{\partial T} = -h_{kl}F_l(T, \mathbf{R}(T, \mathbf{v}))\frac{\partial x_i(t; T, \mathbf{v})}{\partial v_k},$$

(A.32)

can be considered as the linear differential equation; one needs only to supplement it with the corresponding initial condition

$$\mathbf{x}(t; T, \mathbf{v})|_{T=t} = \mathbf{R}(t, \mathbf{v})$$

assuming that function $\mathbf{R}(T, \mathbf{v})$ is known.

The equation for this function can be obtained from the equality

$$\frac{\partial \mathbf{R}(T, \mathbf{v})}{\partial T} = \left.\frac{\partial \mathbf{x}(t; T, \mathbf{v})}{\partial t}\right|_{t=T} + \left.\frac{\partial \mathbf{x}(t; T, \mathbf{v})}{\partial T}\right|_{t=T}.$$

(A.33)

The right-hand side of Eq. (A.33) is the sum of the right-hand sides of Eq. (A.27) and (A.30) at $t = T$. As a result, we obtain the closed nonlinear (quasilinear) equation

$$\frac{\partial \mathbf{R}(T, \mathbf{v})}{\partial T} = -h_{kl} F_l\left(T, \mathbf{R}(T, \mathbf{v})\right) \frac{\partial \mathbf{R}(T, \mathbf{v})}{\partial v_k} + \mathbf{F}\left(T, \mathbf{R}(T, \mathbf{v})\right). \tag{A.34}$$

The initial condition for Eq. (A.34) follows from Eq. (A.28) for $T \to 0$

$$\mathbf{R}(T, \mathbf{v})|_{T=0} = (g + h)^{-1} \mathbf{v}. \tag{A.35}$$

Setting now $t = 0$ in Eq. (A.29), we obtain for the secondary boundary quantity $\mathbf{S}(T, \mathbf{v}) = \mathbf{x}(0; T, \mathbf{v})$ the equation

$$\frac{\partial \mathbf{S}(T, \mathbf{v})}{\partial T} = -h_{kl} F_l\left(T, \mathbf{R}(T, \mathbf{v})\right) \frac{\partial \mathbf{S}(T, \mathbf{v})}{\partial v_k} \tag{A.36}$$

with the initial condition

$$\mathbf{S}(T, \mathbf{v})|_{T=0} = (g + h)^{-1} \mathbf{v}$$

following from Eq. (A.35).

Thus, the problem reduces to the closed quasilinear equation (A.34) with initial value (A.35) and linear equation (A.30) whose coefficients and initial value are determined by the solution of Eq. (A.34).

In the problem under consideration, input 0 and output T are symmetric. For this reason, one can solve it not only from T to 0, but also from 0 to T. In the latter case, functions $\mathbf{R}(T, \mathbf{v})$ and $\mathbf{S}(T, \mathbf{v})$ switch the places.

An important point consists in the fact that, despite the initial problem (A.27) is nonlinear, Eq. (A.30) is the linear equation, because it is essentially the equation in variations. It is Eq. (A.34) that is responsible for nonlinearity.

Note that the above technique of deriving imbedding equations for Eq. (A.27) can be easily extended to the boundary condition of the form

$$\mathbf{g}\left(\mathbf{x}(0)\right) + \mathbf{h}\left(\mathbf{x}(T)\right) + \int_0^T d\tau \mathbf{K}\left(\tau, \mathbf{x}(\tau)\right) = \mathbf{v},$$

where $\mathbf{g}(\mathbf{x})$, $\mathbf{h}(\mathbf{x})$ and $\mathbf{K}(T, \mathbf{x})$ are arbitrary given vector functions.

If function $\mathbf{F}(t, \mathbf{x})$ is linear in $\mathbf{x}$, $F_i(t, \mathbf{x}) = A_{ij}(t) x_j(t)$, then boundary-value problem (A.27), (A.28) assumes the simpler form

$$\frac{d}{dt} \mathbf{x}(t) = A(t) \mathbf{x}(t), \quad g\mathbf{x}(0) + h\mathbf{x}(T) = \mathbf{v},$$

and the solution of Eq. (A.30), (A.34) and (A.36) will be the function linear in $\mathbf{v}$

$$\mathbf{x}(t; T, \mathbf{v}) = X(t; T)\mathbf{v}. \tag{A.37}$$

As a result, we arrive at the closed matrix Riccati equation for matrix $R(T) = X(T; T)$

$$\frac{d}{dT}R(T) = A(T)R(T) - R(T)hA(T)R(T), \quad R(0) = (g + h)^{-1}. \tag{A.38}$$

As regards matrix $X(t, T)$, it satisfies the linear matrix equation with the initial condition

$$\frac{\partial}{\partial T}X(t; T) = -X(t; T)hA(T)R(T), \quad X(t; T)_{T=t} = R(t). \tag{A.39}$$

Note that, for particular linear boundary-value problems of wave propagation, the direct derivation of imbedding equations from the specific problem statement is usually more convenient.

B. Statistical Characteristics of Random Processes, and Fields

B.1 General Remarks

If we deal with random function (random process) $z(t)$, then all statistical characteristics of this function at any fixed instant t are exhaustively described in terms of the one-time probability density

$$P(z, t) = \langle \delta (z(t) - z) \rangle \tag{B.1}$$

dependent parametrically on time t by the following relationship

$$\langle f(z(t)) \rangle = \int\limits_{-\infty}^{\infty} dz f(z) P(z, t).$$

Note that the singular Dirac delta function

$$\varphi(z, t) = \delta (z(t) - z)$$

appeared in Eq. (B.1) in angle brackets of averaging is called the *indicator function*.

The integral distribution function for this process, i.e. the probability of the event that process $z(t) < Z$ at instant t, is calculated by the formula

$$F(t, Z) = \mathsf{P}(z(t) < Z) = \int\limits_{-\infty}^{Z} dz P(z, t)$$

from which follows that

$$F(t, Z) = \langle \theta(Z - z(t)) \rangle, \quad F(t, \infty) = 1, \tag{B.2}$$

where $\theta(z)$ is the Heaviside step function equal to zero for $z < 0$ and unity for $z > 0$. Similar definitions hold for the two-time probability density

$$P(z_1, t_1; z_2, t_2) = \langle \varphi(z_1, t_1; z_2, t_2) \rangle$$

and for the general case of the n-time probability density

$$P(z_1, t_1; \cdots ; z_n, t_n) = \langle \varphi(z_1, t_1; \cdots ; z_n, t_n) \rangle,$$

where

$$\varphi(z_1, t_1; \cdots ; z_n, t_n) = \delta(z(t_1) - z_1) \cdots \delta(z(t_n) - z_n).$$

is the n-time indicator function.

Process $z(t)$ is called *stationary* if all its statistical characteristics are invariant with respect to arbitrary temporal shift, i.e., if

$$P(z_1, t_1 + \tau; \cdots ; z_n, t_n + \tau) = P(z_1, t_1; \cdots ; z_n, t_n). \tag{B.3}$$

In particular, the one-time probability density of stationary process is at all independent of time, and the correlation function depends only on difference of times,

$$B_z(t_1, t_2) = \langle z(t_1)z(t_2) \rangle = B_z(t_1 - t_2).$$

Temporal scale τ_0 characteristic of correlation function $B_z(t)$ is called the temporal correlation radius of process $z(t)$. We can determine this scale, say, by the equality

$$\int\limits_{0}^{\infty} \langle z(t + \tau)z(t) \rangle \, d\tau = \tau_0 \left\langle z^2(t) \right\rangle. \tag{B.4}$$

Note that the Fourier transform of stationary process correlation function

$$\Phi_z(\omega) = \int\limits_{-\infty}^{\infty} dt B_z(t)e^{i\omega t}$$

is called the *temporal spectral function* (or simply *temporal spectrum*).

For random field $f(\mathbf{x}, t)$, the one- and n-point probability densities are defined similarly to those for random processes

$$P(\mathbf{x}, t; f) = \langle \varphi(\mathbf{x}, t; f) \rangle, \tag{B.5}$$

$$P(\mathbf{x}_1, t_1, f_1; \cdots ; \mathbf{x}_n, t_n, f_n) = \langle \varphi(\mathbf{x}_1, t_1, f_1; \cdots ; \mathbf{x}_n, t_n, f_n) \rangle, \tag{B.6}$$

where the indicator functions are defined as follows:

$$\varphi(\mathbf{x}, t; f) = \delta(f(\mathbf{x}, t) - f),$$
$$\varphi(\mathbf{x}_1, t_1, f_1; \cdots ; \mathbf{x}_n, t_n, f_n) = \delta\left(f(\mathbf{x}_1, t_1) - f_1\right) \cdots \delta\left(f(\mathbf{x}_n, t_n) - f_n\right). \tag{B.7}$$

For clarity, we use here variables $\mathbf{x}$ and t as spatial and temporal coordinates; however, in many physical problems, some preferred spatial coordinate can play the role of the temporal coordinate.

Random field $f(\mathbf{x}, t)$ is called the spatially homogeneous field if all its statistical characteristics are invariant relative to spatial translations by arbitrary vector $\mathbf{a}$, i.e., if

$$P(\mathbf{x}_1 + \mathbf{a}, t_1, f_1; \cdots ; \mathbf{x}_n + \mathbf{a}, t_n, f_n) = P(\mathbf{x}_1, t_1, f_1; \cdots ; \mathbf{x}_n, t_n, f_n).$$

In this case, the one-point probability density $P(\mathbf{x}, t; f) = P(t; f)$ is independent of $\mathbf{x}$, and the spatial correlation function $B_f(\mathbf{x}_1, t_1; \mathbf{x}_2, t_2)$ depends on the difference $\mathbf{x}_1 - \mathbf{x}_2$:

$$B_f(\mathbf{x}_1, t_1; \mathbf{x}_2, t_2) = \langle f(\mathbf{x}_1, t_1) f(\mathbf{x}_2, t_2) \rangle = B_f(\mathbf{x}_1 - \mathbf{x}_2; t_1, t_2).$$

If random field $f(\mathbf{x}, t)$ is additionally invariant with respect to rotation of all vectors $\mathbf{x}_i$ by arbitrary angle, i.e., with respect to rotations of the reference system, then field $f(\mathbf{x}, t)$ is called the homogeneous isotropic random field. In this case, the correlation function depends only on length $|\mathbf{x}_1 - \mathbf{x}_2|$:

$$B_f(\mathbf{x}_1, t_1; \mathbf{x}_2, t_2) = \langle f(\mathbf{x}_1, t_1) f(\mathbf{x}_2, t_2) \rangle = B_f(|\mathbf{x}_1 - \mathbf{x}_2|; t_1, t_2).$$

The corresponding Fourier transform of the correlation function with respect to the spatial variable defines the spatial spectral function (called also the angular spectrum)

$$\Phi_f(\mathbf{k}, t) = \int d\mathbf{x} B_f(\mathbf{x}, t) e^{i\mathbf{k}\mathbf{x}},$$

and the Fourier transform of the correlation function of random field $f(\mathbf{x}, t)$ stationary in time and homogeneous in space defines the space–time spectrum

$$\Phi_f(\mathbf{k}, \omega) = \int d\mathbf{x} \int\limits_{-\infty}^{\infty} dt B_f(\mathbf{x}, t) e^{i(\mathbf{k}\mathbf{x}+\omega t)}.$$

In the case of isotropic random field $f(\mathbf{x}, t)$, the space–time spectrum appears isotropic in the $\mathbf{k}$-space:

$$\Phi_f(\mathbf{k}, \omega) = \Phi_f(k, \omega).$$

An exhaustive description of random function $z(t)$ can be given in terms of the characteristic functional

$$\Phi[v(\tau)] = \left\langle \exp\left\{ i \int\limits_{-\infty}^{\infty} d\tau v(\tau)z(\tau) \right\} \right\rangle,$$

where $v(t)$ is arbitrary (but sufficiently smooth) function. Functional $\Phi[v(\tau)]$ being known, one can determine such characteristics of random function $z(t)$ as mean value $\langle z(t) \rangle$, correlation function $\langle z(t_1)z(t_2) \rangle$, n-time moment function $\langle z(t_1) \cdots z(t_n) \rangle$, etc.

Indeed, expanding functional $\Phi[v(\tau)]$ in the functional Taylor series, we obtain the representation of characteristic functional in terms of the moment functions of process $z(t)$:

$$\Phi[v(\tau)] = \sum_{n=0}^{\infty} \frac{i^n}{n!} \int\limits_{-\infty}^{\infty} dt_1 \cdots \int\limits_{-\infty}^{\infty} dt_n M_n(t_1, \ldots, t_n)v(t_1) \cdots v(t_n),$$

$$M_n(t_1, \ldots, t_n) = \langle z(t_1) \cdots z(t_n) \rangle = \frac{1}{i^n} \frac{\delta^n}{\delta v(t_1) \cdots \delta v(t_n)} \Phi[v(\tau)] \bigg|_{v=0}.$$

Consequently, the moment functions of random process $z(t)$ are expressed in terms of variational derivatives of the characteristic functional.

Represent functional $\Phi[v(\tau)]$ in the form $\Phi[v(\tau)] = \exp\{\Theta[v(\tau)]\}$. Functional $\Theta[v(\tau)]$ also can be expanded in the functional Taylor series

$$\Theta[v(\tau)] = \sum_{n=1}^{\infty} \frac{i^n}{n!} \int\limits_{-\infty}^{\infty} dt_1 \cdots \int\limits_{-\infty}^{\infty} dt_n K_n(t_1, \ldots, t_n)v(t_1) \cdots v(t_n), \qquad \text{(B.8)}$$

where function

$$K_n(t_1, \ldots, t_n) = \frac{1}{i^n} \frac{\delta^n}{\delta v(t_1) \cdots \delta v(t_n)} \Theta[v(\tau)] \bigg|_{v=0}$$

is called the n-th order *cumulant function* of random process $z(t)$.

The characteristic functional and the n-th order cumulant functions of scalar random field $f(\mathbf{x}, t)$ are defined similarly

$$\Phi[v(\mathbf{x}', \tau)] = \left\langle \exp\left\{ i \int d\mathbf{x} \int\limits_{-\infty}^{\infty} dt v(\mathbf{x}, t) f(\mathbf{x}, t) \right\} \right\rangle = \exp\left\{ \Theta[v(\mathbf{x}', \tau)] \right\},$$

$$M_n(\mathbf{x}_1, t_1, \ldots, \mathbf{x}_n, t_n) = \frac{1}{i^n} \frac{\delta^n}{\delta v(\mathbf{x}_1, t_1) \cdots \delta v(\mathbf{x}_n, t_n)} \Phi[v(\mathbf{x}', \tau)] \Big|_{v=0},$$

$$K_n(\mathbf{x}_1, t_1, \ldots, \mathbf{x}_n, t_n) = \frac{1}{i^n} \frac{\delta^n}{\delta v(\mathbf{x}_1, t_1) \cdots \delta v(\mathbf{x}_n, t_n)} \Theta[v(\mathbf{x}', \tau)] \Big|_{v=0}.$$

In the case of vector random field $\mathbf{f}(\mathbf{x}, t)$, we must assume that $\mathbf{v}(\mathbf{x}, t)$ is the vector function.

As was mentioned, characteristic functionals ensure a complete description of random processes and fields. However, even one-point probability densities of random processes and fields give certain data on random process evolution in the entire interval of times and on the spatial structure of random fields, as well. These data can be obtained on the basis of ideas of statistical topography of random processes and fields.

B.2 Gaussian Random Process

We start the discussion with the continuous processes; namely, we consider the Gaussian random process $z(t)$ with zero-valued mean ($\langle z(t) \rangle = 0$) and correlation function $B(t_1, t_2) = \langle z(t_1)z(t_2) \rangle$. The corresponding characteristic functional assumes the form

$$\Phi[v(\tau)] = \exp\left\{ -\frac{1}{2} \int\limits_{-\infty}^{\infty} \int\limits_{-\infty}^{\infty} dt_1 dt_2 B(t_1, t_2) v(t_1) v(t_2) \right\}. \tag{B.9}$$

Only one cumulant function (the correlation function)

$$K_2(t_1, t_2) = B(t_1, t_2)$$

is different from zero for this process, so that

$$\Theta[v(\tau)] = -\frac{1}{2} \int\limits_{-\infty}^{\infty} \int\limits_{-\infty}^{\infty} dt_1 dt_2 B(t_1, t_2) v(t_1) v(t_2). \tag{B.10}$$

Consider the nth-order variational derivative of functional $\Phi[v(\tau)]$. It satisfies the following line of equalities:

$$\frac{\delta^n}{\delta v(t_1)\cdots\delta v(t_n)}\Phi[v(\tau)] = \frac{\delta^{n-1}}{\delta v(t_2)\cdots\delta v(t_n)}\frac{\delta\Theta[v(\tau)]}{\delta v(t_1)}\Phi[v(\tau)]$$

$$= \frac{\delta^2\Theta[v(\tau)]}{\delta v(t_1)\delta v(t_2)}\frac{\delta^{n-2}}{\delta v(t_3)\cdots\delta v(t_n)}\Phi[v(\tau)] + \frac{\delta^{n-2}}{\delta v(t_3)\cdots\delta v(t_n)}\frac{\delta\Theta[v(\tau)]}{\delta v(t_1)}\frac{\delta\Phi[v(\tau)]}{\delta v(t_2)}.$$

Setting now $v = 0$, we obtain that moment functions of the Gaussian process $z(t)$ satisfy the recurrence formula

$$M_n(t_1,\ldots,t_n) = \sum_{k=2}^n B(t_1,t_2)M_{n-2}(t_2,\ldots,t_{k-1},t_{k+1},\ldots,t_n). \qquad \text{(B.11)}$$

From this formula follows that, for the Gaussian process with zero-valued mean, all moment functions of odd orders are identically equal to zero and the moment functions of even orders are represented as sums of terms which are the products of averages of all possible pairs $z(t_i)z(t_k)$.

If we assume that function $v(\tau)$ in Eq. (B.10) is different from zero only in interval $0 < \tau < t$, the characteristic functional

$$\Phi[t;v(\tau)] = \left\langle\exp\left(i\int_0^t d\tau z(\tau)v(\tau)\right)\right\rangle = \exp\left\{-\int_0^t d\tau_1\int_0^{\tau_1}d\tau_2 B(\tau_1,\tau_2)v(\tau_1)v(\tau_2)\right\}$$

(B.12)

becomes a function of time t and satisfies the ordinary differential equation

$$\frac{d}{dt}\Phi[t;v(\tau)] = -v(t)\int_0^t dt_1 B(t,t_1)v(t_1)\Phi[t;v(\tau)], \quad \Phi[0;v(\tau)] = 1. \quad \text{(B.13)}$$

To obtain the one-time characteristic function of the Gaussian random process at instant t, we specify function $v(\tau)$ in Eq. (B.9) in the form

$$v(\tau) = v\delta(\tau - t).$$

Then, we obtain

$$\Phi(v,t) = \left\langle e^{ivz(t)}\right\rangle = \int_{-\infty}^\infty dz\, P(z,t)e^{ivz} = \exp\left\{-\frac{1}{2}\sigma^2(t)v^2\right\}, \qquad \text{(B.14)}$$

where $\sigma^2(t) = B(t,t)$. Using the inverse Fourier transform of (B.14), we obtain the one-time probability density of the Gaussian random process

$$P(z, t) = \frac{1}{2\pi} \int\limits_{-\infty}^{\infty} dv\, \Phi(v, t) e^{-ivz} = \frac{1}{\sqrt{2\pi\sigma^2(t)}} \exp\left\{-\frac{z^2}{2\sigma^2(t)}\right\}. \qquad (B.15)$$

Note that, in the case of stationary random process $z(t)$, variance $\sigma^2(t)$ is independent of time t, i.e., $\sigma^2(t) = \sigma^2 = \text{const}$.

Density $P(z, t)$ as a function of z is symmetric relative to point $z = 0$,

$$P(z, t) = P(-z, t).$$

If mean value of the Gaussian random process is different from zero, then we can consider process $z(t) - \langle z(t) \rangle$ to obtain instead of Eq. (B.15) the expression

$$P(z, t) = \frac{1}{\sqrt{2\pi\sigma^2(t)}} \exp\left\{-\frac{(z - \langle z(t)\rangle)^2}{2\sigma^2(t)}\right\} \qquad (B.16)$$

and the corresponding integral distribution function assumes the form

$$F(z, t) = \frac{1}{\sqrt{2\pi\sigma^2(t)}} \int\limits_{-\infty}^{z} dz\, \exp\left\{-\frac{(z - \langle z(t)\rangle)^2}{2\sigma^2(t)}\right\} = \text{Pr}\left(\frac{z - \langle z(t)\rangle}{\sigma(t)}\right),$$

where probability integral $\text{Pr}(z)$ is defined by Eq. (5.4).

B.3 Correlation Splitting for Random Processes and Fields

For simplicity, we content ourselves here with the one-dimensional random processes (extensions to multidimensional cases are obvious). We need the ability of calculating correlation $\langle z(t)]R[z(\tau)]\rangle$, where $R[z(\tau)]$ is the functional that can depend on process $z(t)$ both explicitly and implicitly.

To calculate this average, we consider auxiliary functional $R[z(\tau) + \eta(\tau)]$, where $\eta(t)$ is arbitrary deterministic function, and calculate the correlation

$$\langle z(t)R[z(\tau) + \eta(\tau)]\rangle. \qquad (B.17)$$

The correlation of interest will be obtained by setting $\eta(\tau) = 0$ in the final result.

We can expand the above auxiliary functional $R[z(\tau) + \eta(\tau)]$ in the functional Taylor series with respect to $z(\tau)$. The result can be represented in the form

$$R[z(\tau) + \eta(\tau)] = \exp\left\{\int\limits_{-\infty}^{\infty} d\tau z(\tau) \frac{\delta}{\delta\eta(\tau)}\right\} R[\eta(\tau)],$$

where we introduced the functional shift operator. With this representation, we can obtain the following expression for correlation (B.17)

$$\langle z(t)R[z(\tau) + \eta(\tau)]\rangle = \Omega\left[t; \frac{\delta}{i\delta\eta(\tau)}\right]\langle R[z(\tau) + \eta(\tau)]\rangle, \tag{B.18}$$

where functional

$$\Omega[t; v(\tau)] = \frac{\left\langle z(t)\exp\left\{i\int\limits_{-\infty}^{\infty} d\tau z(\tau)v(\tau)\right\}\right\rangle}{\left\langle \exp\left\{i\int\limits_{-\infty}^{\infty} d\tau z(\tau)v(\tau)\right\}\right\rangle}$$

$$= \frac{1}{\Phi[v(\tau)]}\frac{\delta}{i\delta v(t)}\Phi[v(\tau)] = \frac{\delta}{i\delta v(t)}\Theta[v(\tau)]. \tag{B.19}$$

Here $\Theta[v(\tau)] = \ln\Phi[v(\tau)]$ and $\Phi[v(\tau)]$ is the characteristic functional of random process $z(t)$.

Replacing now differentiation with respect to $\eta(\tau)$ by differentiation with respect to $z(\tau)$ and setting $\eta(\tau) = 0$, we obtain the expression

$$\langle z(t)R[z(\tau)]\rangle = \left\langle \Omega\left[t; \frac{\delta}{i\delta z(\tau)}\right]R[z(\tau)]\right\rangle. \tag{B.20}$$

If we expand functional $\Theta[v(\tau)]$ in the functional Taylor series (B.8) and differentiate the result with respect to $v(t)$, we obtain

$$\Omega[t; v(\tau)] = \sum_{n=0}^{\infty} \frac{i^n}{n!}\int\limits_{-\infty}^{\infty} dt_1 \cdots \int\limits_{-\infty}^{\infty} dt_n K_{n+1}(t, t_1, \ldots, t_n)v(t_1)\cdots v(t_n)$$

and expression (B.20) will assume the form

$$\langle z(t)R[z(\tau)]\rangle = \sum_{n=0}^{\infty} \frac{1}{n!}\int\limits_{-\infty}^{\infty} dt_1 \cdots \int\limits_{-\infty}^{\infty} dt_n K_{n+1}(t, t_1, \ldots, t_n)\left\langle \frac{\delta^n R[z(\tau)]}{\delta z(t_1)\cdots\delta z(t_n)}\right\rangle. \tag{B.21}$$

In physical problems satisfying the condition of dynamic causality in time, statistical characteristics of the solution at instant t depend on the statistical characteristics of process $z(\tau)$ for $0 \leq \tau \leq t$, which are completely described by the characteristic functional

$$\Phi[t; v(\tau)] = \exp\{\Theta[t; v(\tau)]\} = \left\langle \exp\left\{i\int\limits_{0}^{t} d\tau z(\tau)v(\tau)\right\}\right\rangle.$$

In this case, the obtained formulas hold also for calculating statistical averages $\langle z(t')R[t; z(\tau)]\rangle$ for $t' < t$, $\tau \le t$, i.e., we have the equality

$$\langle z(t')R[t; z(\tau)]\rangle = \left\langle \Omega\left[t', t; \frac{\delta}{i\delta z(\tau)}\right] R[t; z(\tau)]\right\rangle \quad (0 < t' < t), \quad (B.22)$$

where

$$\Omega[t', t; v(\tau)] = \frac{\delta}{i\delta v(t')}\Theta[t; v(\tau)]$$

$$= \sum_{n=0}^{\infty} \frac{i^n}{n!} \int_0^t dt_1 \cdots \int_0^t dt_n K_{n+1}(t', t_1, \ldots, t_n)v(t_1)\cdots v(t_n). \quad (B.23)$$

For $t' = t - 0$, formula (B.22) holds as before, i.e.

$$\langle z(t)R[t; z(\tau)]\rangle = \left\langle \Omega\left[t, t; \frac{\delta}{i\delta z(\tau)}\right] R[t; z(\tau)]\right\rangle. \quad (B.24)$$

However, expansion (B.23) not always gives the correct result in the limit $t' \rightarrow t - 0$ (which means that the limiting process and the procedure of expansion in the functional Taylor series can be non-commutable). In this case,

$$\Omega[t, t; v(\tau)] = \frac{\left\langle z(t)\exp\left\{i\int_0^t d\tau z(\tau)v(\tau)\right\}\right\rangle}{\left\langle \exp\left\{i\int_0^t d\tau z(\tau)v(\tau)\right\}\right\rangle} = \frac{d}{iv(t)dt}\Theta[t; v(\tau)], \quad (B.25)$$

and statistical averages in Eqs. (B.22) and (B.24) can be discontinuous at $t' = t - 0$.

B.3.1 Correlation Splitting for Random Gaussian Processes and Fields (Furutzu–Novikov Formula)

In the case of the Gaussian random process $z(t)$, all formulas obtained in the previous section become significantly simpler. In this case, the logarithm of characteristic functional $\Phi[v(\tau)]$ is given by Eq. (B.10) (we assume that the mean value of process $z(t)$ is zero), and functional $\Theta[t, v(\tau)]$ assumes the form

$$\Theta[t, v(\tau)] = -\frac{1}{2}\int_{-\infty}^{\infty}\int_{-\infty}^{\infty} d\tau_1 d\tau_2 B(\tau_1, \tau_2)v(\tau_1)v(\tau_2).$$

As a consequence, functional $\Omega[t; v(\tau)]$ (B.19) is the linear functional

$$\Omega[t; v(\tau)] = i \int_{-\infty}^{\infty} d\tau_1 B(t, \tau_1) v(\tau_1), \tag{B.26}$$

and Eq. (B.18) assumes the form

$$\langle z(t) R[z(\tau) + \eta(\tau)] \rangle = \int_{-\infty}^{\infty} d\tau_1 B(t, \tau_1) \frac{\delta}{\delta \eta(\tau_1)} \langle R[z(\tau) + \eta(\tau)] \rangle . \tag{B.27}$$

Replacing differentiation with respect to $\eta(\tau)$ by differentiation with respect to $z(\tau)$ and setting $\eta(\tau) = 0$, we obtain the equality

$$\langle z(t) R[z(\tau)] \rangle = \int_{-\infty}^{\infty} d\tau_1 B(t, \tau_1) \left\langle \frac{\delta}{\delta z(\tau_1)} R[z(\tau)] \right\rangle \tag{B.28}$$

commonly known in physics as the *Furutsu–Novikov formula*.

One can easily obtain the multi-dimensional extension of Eq. (B.28); it can be written in the form

$$\langle z_{i_1, \ldots, i_n}(\mathbf{r}) R[\mathbf{z}] \rangle = \int d\mathbf{r}' \langle z_{i_1, \ldots, i_n}(\mathbf{r}) z_{j_1, \ldots, j_n}(\mathbf{r}') \rangle \left\langle \frac{\delta R[\mathbf{z}]}{\delta z_{j_1, \ldots, j_n}(\mathbf{r}')} \right\rangle, \tag{B.29}$$

where $\mathbf{r}$ stands for all continuous arguments of random vector field $\mathbf{z}(\mathbf{r})$ and $i_1, \ldots, i_n$ are the discrete (index) arguments. Repeated index arguments in the right-hand side of Eq. (B.29) assume summation.

If random process $z(\tau)$ is defined only on time interval $[0, t]$, then functional $\Theta[t, v(\tau)]$ assumes the form

$$\Theta[t, v(\tau)] = -\frac{1}{2} \int_0^t \int_0^t d\tau_1 d\tau_2 B(\tau_1, \tau_2) v(\tau_1) v(\tau_2), \tag{B.30}$$

and functionals $\Omega[t', t; v(\tau)]$ and $\Omega[t, t; v(\tau)]$ are the linear functionals

$$\Omega[t', t; v(\tau)] = \frac{\delta}{i \delta v(t')} \Theta[t, v(\tau)] = i \int_0^t d\tau B(t', \tau) v(\tau),$$

$$\Omega[t, t; v(\tau)] = \frac{d}{iv(t)dt} \Theta[t, v(\tau)] = i \int_0^t d\tau B(t, \tau) v(\tau). \tag{B.31}$$

As a consequence, Eqs. (B.22), (B.24) assume the form

$$\langle z(t')R[t, z(\tau)]\rangle = \int_0^t d\tau B(t', \tau)\left\langle \frac{\delta R[z(\tau)]}{\delta z(\tau)}\right\rangle \quad (t' \leqslant t) \tag{B.32}$$

that coincides with Eq. (B.28) if the condition

$$\frac{\delta R[t; z(\tau)]}{\delta z(\tau)} = 0 \quad \text{for} \quad \tau < 0, \quad \tau > t \tag{B.33}$$

holds.

C. Approximation of Gaussian Random Field Delta-Correlated in Time

C.1 The Fokker–Planck Equation

Let vector function $\mathbf{x}(t) = \{x_1(t), x_2(t), \ldots, x_n(t)\}$ satisfies the dynamic equation

$$\frac{d}{dt}\mathbf{x}(t) = \mathbf{v}(\mathbf{x}, t) + \mathbf{f}(\mathbf{x}, t), \quad \mathbf{x}(t_0) = \mathbf{x}_0, \tag{C.1}$$

where $v_i(\mathbf{x}, t)$ $(i = 1, \ldots, n)$ are the deterministic functions and $f_i(\mathbf{x}, t)$ are the random functions of $(n + 1)$ variable that have the following properties:

(a) $f_i(\mathbf{x}, t)$ is the Gaussian random field in the $(n + 1)$-dimensional space $(\mathbf{x}, t)$;
(b) $\langle f_i(\mathbf{x}, t)\rangle = 0$.

For definiteness, we assume that t is the temporal variable and $\mathbf{x}$ is the spatial variable.

Statistical characteristics of field $f_i(\mathbf{x}, t)$ are completely described by correlation tensor

$$B_{ij}(\mathbf{x}, t; \mathbf{x}', t') = \langle f_i(\mathbf{x}, t)f_j(\mathbf{x}', t')\rangle.$$

Because Eq. (C.1) is the first-order equation with the initial value, its solution satisfies the dynamic causality condition

$$\frac{\delta}{\delta f_j(\mathbf{x}', t')}x_i(t) = 0 \quad \text{for} \quad t' < t_0 \quad \text{and} \quad t' > t, \tag{C.2}$$

which means that solution $\mathbf{x}(t)$ depends only on values of function $f_j(\mathbf{x}, t')$ for times t' preceding time t, i.e., $t_0 \leq t' \leq t$. In addition, we have the following equality for the variational derivative

$$\frac{\delta}{\delta f_j(\mathbf{x}', t - 0)} x_i(t) = \delta_{ij}\delta(\mathbf{x}(t) - \mathbf{x}').\tag{C.3}$$

Nevertheless, the statistical relationship between $\mathbf{x}(t)$ and function values $f_j(\mathbf{x}, t'')$ for consequent times $t'' > t$ can exist, because such function values $f_j(\mathbf{x}, t'')$ are correlated with values $f_j(\mathbf{x}, t')$ for $t' \leq t$. It is obvious that the correlation between function $\mathbf{x}(t)$ and consequent values $f_j(\mathbf{x}, t'')$ is appreciable only for $t'' - t \leq \tau_0$, where τ_0 is the correlation radius of field $\mathbf{f}(\mathbf{x}, t)$ with respect to variable t.

For many actual physical processes, characteristic temporal scale T of function $\mathbf{x}(t)$ significantly exceeds correlation radius τ_0 ($T \gg \tau_0$); in this case, the problem has small parameter τ_0/T that can be used to construct an approximate solution.

In the first approximation with respect to this small parameter, one can consider the asymptotic solution for $\tau_0 \to 0$. In this case values of function $\mathbf{x}(t')$ for $t' < t$ will be independent of values $\mathbf{f}(\mathbf{x}, t'')$ for $t'' > t$ not only functionally, but also statistically. This approximation is equivalent to the replacement of correlation tensor B_{ij} with the effective tensor

$$B_{ij}^{\text{eff}}(\mathbf{x}, t; \mathbf{x}', t') = 2\delta(t - t')F_{ij}(\mathbf{x}, \mathbf{x}'; t).\tag{C.4}$$

Here, quantity $F_{ij}(\mathbf{x}, \mathbf{x}', t)$ is determined from the condition that integrals of $B_{ij}(\mathbf{x}, t; \mathbf{x}', t')$ and $B_{ij}^{\text{eff}}(\mathbf{x}, t; \mathbf{x}', t')$ over t' coincide

$$F_{ij}(\mathbf{x}, \mathbf{x}', t) = \frac{1}{2}\int\limits_{-\infty}^{\infty} dt' B_{ij}(\mathbf{x}, t; \mathbf{x}', t'),$$

which just corresponds to the passage to the Gaussian random field delta-correlated in time t.

Introduce the indicator function

$$\varphi(\mathbf{x}, t) = \delta(\mathbf{x}(t) - \mathbf{x}),\tag{C.5}$$

where $\mathbf{x}(t)$ is the solution to Eq. (C.1), which satisfies the Liouville equation

$$\left(\frac{\partial}{\partial t} + \frac{\partial}{\partial \mathbf{x}}\mathbf{v}(\mathbf{x}, t)\right)\varphi(\mathbf{x}, t) = -\frac{\partial}{\partial \mathbf{x}}\mathbf{f}(\mathbf{x}, t)\varphi(\mathbf{x}, t)\tag{C.6}$$

and the equality

$$\frac{\delta}{\delta f_j(\mathbf{x}', t - 0)}\varphi(\mathbf{x}, t) = -\frac{\partial}{\partial x_j}\left\{\delta(\mathbf{x} - \mathbf{x}')\varphi(\mathbf{x}, t)\right\}.\tag{C.7}$$

The equation for the probability density of the solution to Eq. (C.1)

$$P(\mathbf{x}, t) = \langle\varphi(\mathbf{x}, t)\rangle = \langle\delta(\mathbf{x}(t) - \mathbf{x})\rangle$$

can be obtained by averaging Eq. (C.6) over an ensemble of realizations of field $\mathbf{f}(\mathbf{x}, t)$,

$$\left(\frac{\partial}{\partial t} + \frac{\partial}{\partial \mathbf{x}} \mathbf{v}(\mathbf{x}, t)\right) P(\mathbf{x}, t) = -\frac{\partial}{\partial \mathbf{x}} \langle \mathbf{f}(\mathbf{x}, t)\varphi(\mathbf{x}, t)\rangle . \tag{C.8}$$

We rewrite Eq. (C.8) in the form

$$\left(\frac{\partial}{\partial t} + \frac{\partial}{\partial \mathbf{x}} \mathbf{v}(\mathbf{x}, t)\right) P(\mathbf{x}, t)$$

$$= -\frac{\partial}{\partial x_i} \int d\mathbf{x}' \int_{t_0}^{t} dt' B_{ij}(\mathbf{x}, t; \mathbf{x}', t') \left\langle \frac{\delta \varphi(\mathbf{x}, t)}{\delta f_j(\mathbf{x}', t')} \right\rangle , \tag{C.9}$$

where we used the Furutsu–Novikov formula (B.29)

$$\langle f_k(\mathbf{x}, t) R[t; \mathbf{f}(\mathbf{y}, \tau)]\rangle = \int d\mathbf{x}' \int dt' B_{kl}(\mathbf{x}, t; \mathbf{x}', t') \left\langle \frac{\delta R[t; \mathbf{f}(\mathbf{y}, \tau)]}{\delta f_l(\mathbf{x}', t')} \right\rangle \tag{C.10}$$

for the correlator of the Gaussian random field $\mathbf{f}(\mathbf{x}, t)$ with arbitrary functional $R[t; \mathbf{f}(\mathbf{y}, \tau)]$ of it and the dynamic causality condition (C.2).

Equation (C.9) shows that the one-time probability density of solution $\mathbf{x}(t)$ at instant t is governed by functional dependence of solution $\mathbf{x}(t)$ on field $\mathbf{f}(\mathbf{x}', t)$ for all times in the interval (t_0, t).

In the general case, there is no closed equation for the probability density $P(\mathbf{x}, t)$. However, if we use approximation (C.4) for the correlation function of field $\mathbf{f}(\mathbf{x}, t)$, there appear terms related to variational derivatives $\delta \varphi[\mathbf{x}, t; \mathbf{f}(\mathbf{y}, \tau)]/\delta f_j(\mathbf{x}', t')$ at coincident temporal arguments $t' = t - 0$,

$$\left(\frac{\partial}{\partial t} + \frac{\partial}{\partial \mathbf{x}} \mathbf{v}(\mathbf{x}, t)\right) P(\mathbf{x}, t) = -\frac{\partial}{\partial x_i} \int d\mathbf{x}' F_{ij}(\mathbf{x}, \mathbf{x}', t) \left\langle \frac{\delta \varphi(\mathbf{x}, t)}{\delta f_j(\mathbf{x}', t - 0)} \right\rangle .$$

According to Eq. (C.7), these variational derivatives can be expressed immediately in terms of quantity $\varphi[\mathbf{x}, t; \mathbf{f}(\mathbf{y}, \tau)]$. Thus, we obtain the closed Fokker–Planck equation

$$\left(\frac{\partial}{\partial t} + \frac{\partial}{\partial x_k} [v_k(\mathbf{x}, t) + A_k(\mathbf{x}, t)]\right) P(\mathbf{x}, t) = \frac{\partial^2}{\partial x_k \partial x_l} [F_{kl}(\mathbf{x}, \mathbf{x}, t) P(\mathbf{x}, t)] , \tag{C.11}$$

where

$$A_k(\mathbf{x}, t) = \frac{\partial}{\partial x_l'} F_{kl}(\mathbf{x}, \mathbf{x}', t) \bigg|_{\mathbf{x}'=\mathbf{x}} .$$

Equation (C.11) should be solved with the initial condition

$$P(\mathbf{x}, t_0) = \delta(\mathbf{x} - \mathbf{x}_0),$$

or with a more general initial condition $P(\mathbf{x}, t_0) = W_0(\mathbf{x})$ if the initial conditions are also random, but statistically independent of field $\mathbf{f}(\mathbf{x}, t)$.

The Fokker–Planck equation (C.11) is a partial differential equation and its further analysis essentially depends on boundary conditions with respect to $\mathbf{x}$ whose form can vary depending on the problem at hand.

Consider the quantities appeared in Eq. (C.11). In this equation, the terms containing $A_k(\mathbf{x}, t)$ and $F_{kl}(\mathbf{x}, \mathbf{x}', t)$ are stipulated by fluctuations of field $\mathbf{f}(\mathbf{x}, t)$. If field $\mathbf{f}(\mathbf{x}, t)$ is stationary in time, quantities $A_k(\mathbf{x})$ and $F_{kl}(\mathbf{x}, \mathbf{x}')$ are independent of time. If field $\mathbf{f}(\mathbf{x}, t)$ is additionally homogeneous and isotropic in all spatial coordinates, then

$$F_{kl}(\mathbf{x}, \mathbf{x}, t) = \text{const},$$

which corresponds to the constant tensor of diffusion coefficients, and $A_k(\mathbf{x}, t) = 0$ (note however that quantities $F_{kl}(\mathbf{x}, \mathbf{x}', t)$ and $A_k(\mathbf{x}, t)$ can depend on $\mathbf{x}$ because of the use of a curvilinear coordinate systems).

C.2 Transition Probability Distributions

Turn back to dynamic system (C.1) and consider the m-time probability density

$$P_m(\mathbf{x}_1, t_1; \cdots; \mathbf{x}_m, t_m) = \langle \delta(\mathbf{x}(t_1) - \mathbf{x}_1) \cdots \delta(\mathbf{x}(t_m) - \mathbf{x}_m) \rangle \quad (C.12)$$

for m different instants $t_1 < t_2 < \cdots < t_m$. Differentiating Eq. (C.12) with respect to time t_m and using then dynamic equation (C.1), dynamic causality condition (C.2), definition of function $F_{kl}(\mathbf{x}, \mathbf{x}', t)$, and the Furutsu–Novikov formula (C.10), one can obtain the equation similar to the Fokker–Planck equation (C.11),

$$\frac{\partial}{\partial t_m} P_m(\mathbf{x}_1, t_1; \cdots; \mathbf{x}_m, t_m)$$

$$+ \sum_{k=1}^{n} \frac{\partial}{\partial x_{mk}} [v_k(\mathbf{x}_m, t_m) + A_k(\mathbf{x}_m, t_m)] P_m(\mathbf{x}_1, t_1; \cdots; \mathbf{x}_m, t_m)$$

$$= \sum_{k=1}^{n} \sum_{l=1}^{n} \frac{\partial^2}{\partial x_{mk} \partial x_{ml}} [F_{kl}(\mathbf{x}_m, \mathbf{x}_m, t_m) P_m(\mathbf{x}_1, t_1; \cdots; \mathbf{x}_m, t_m))] . \quad (C.13)$$

No summation over index m is performed here. The initial value to Eq. (C.13) can be determined from Eq. (C.12). Setting $t_m = t_{m-1}$ in (C.12), we obtain

$$P_m(\mathbf{x}_1, t_1; \cdots; \mathbf{x}_m, t_{m-1})$$

$$= \delta(\mathbf{x}_m - \mathbf{x}_{m-1}) P_{m-1}(\mathbf{x}_1, t_1; \cdots; \mathbf{x}_{m-1}, t_{m-1}). \quad (C.14)$$

Equation (C.13) assumes the solution in the form

$$P_m(\mathbf{x}_1, t_1; \cdots; \mathbf{x}_m, t_m)$$
$$= p(\mathbf{x}_m, t_m|\mathbf{x}_{m-1}, t_{m-1})P_{m-1}(\mathbf{x}_1, t_1; \cdots; \mathbf{x}_{m-1}, t_{m-1}). \quad \text{(C.15)}$$

Because all differential operations in Eq. (C.13) concern only t_m and $\mathbf{x}_m$, we can find the equation for the *transitional probability density* by substituting Eq. (C.15) in (C.13) and (C.14):

$$\left(\frac{\partial}{\partial t} + \frac{\partial}{\partial x_k}[v_k(\mathbf{x}, t) + A_k(\mathbf{x}, t)]\right)p(\mathbf{x}, t|\mathbf{x}_0, t_0)$$

$$= \frac{\partial^2}{\partial x_k \partial x_l}\left[F_{kl}(\mathbf{x}, \mathbf{x}, t)p(\mathbf{x}, t|\mathbf{x}_0, t_0)\right] \quad \text{(C.16)}$$

with initial condition

$$p(\mathbf{x}, t|\mathbf{x}_0, t_0)|_{t \to t_0} = \delta(\mathbf{x} - \mathbf{x}_0),$$

where

$$p(\mathbf{x}, t|\mathbf{x}_0, t_0) = \langle \delta(\mathbf{x}(t) - \mathbf{x})|\mathbf{x}(t_0) = \mathbf{x}_0 \rangle.$$

In Eq. (C.16) we denoted variables $\mathbf{x}_m$ and t_m as $\mathbf{x}$ and t, and variables $\mathbf{x}_{m-1}$ and t_{m-1} as $\mathbf{x}_0$ and t_0.

Using formula (C.15) $(m - 1)$ times, we obtain the relationship

$$P_m(\mathbf{x}_1, t_1; \cdots; \mathbf{x}_m, t_m)$$
$$= p(\mathbf{x}_m, t_m|\mathbf{x}_{m-1}, t_{m-1}) \cdots p(\mathbf{x}_2, t_2|\mathbf{x}_1, t_1)P(\mathbf{x}_1, t_1), \quad \text{(C.17)}$$

where $P(\mathbf{x}_1, t_1)$ is the one-time probability density governed by Eq. (C.11). Equality (C.17) expresses the multi-time probability density as the product of transitional probability densities, which means that random process $\mathbf{x}(t)$ is the Markovian process.

Equation (C.11) is usually called the *forward Fokker–Planck equation*. The *backward Fokker–Planck equation* (it describes the transitional probability density as a function of the initial parameters t_0 and $\mathbf{x}_0$) can also be easily derived (see, for example, monographs [28, 29]).

C.3 Applicability Range of the Fokker–Planck Equation

To estimate the applicability range of the Fokker–Planck equation, we must include into consideration the finite-valued correlation radius τ_0 of field $\mathbf{f}(\mathbf{x}, t)$ with respect to time. Thus, smallness of parameter τ_0/T is the necessary but generally not sufficient condition in order that one can describe statistical characteristics of the solution to Eq. (C.1) using the approximation of the delta-correlated random field of which a consequence is the Fokker–Planck equation. Every particular problem requires more detailed investigations. Below, we give a more physical method called the *diffusion approximation*. This method also leads to the Markovian property of the solution to Eq. (C.1); however, it considers to some extent the finite value of the temporal correlation radius.

Here, we emphasize that the approximation of the delta-correlated random field does not reduce to the formal replacement of random field $\mathbf{f}(\mathbf{x}, t)$ in Eq. (C.1) with the random field with correlation function (C.4). This approximation corresponds to the construction of an asymptotic expansion in temporal correlation radius τ_0 of filed $\mathbf{f}(\mathbf{x}, t)$ for $\tau_0 \to 0$. It is in such limit process that exact average quantities like

$$\langle \mathbf{f}(\mathbf{x}, t) R[t; \mathbf{f}(\mathbf{x}', \tau)] \rangle$$

grade into the expressions obtained by the formal replacement of the correlation tensor of field $\mathbf{f}(\mathbf{x}, t)$ with the effective tensor (C.4).

C.3.1 Langevin Equation

We illustrate the above speculation by the example of the *Langevin equation* that allows an exhaustive statistical analysis. This equation has the form

$$\frac{d}{dt} x(t) = -\lambda x(t) + f(t), \quad x(t_0) = 0 \tag{C.18}$$

and assumes that the sufficiently fine smooth function $f(t)$ is the stationary Gaussian process with zero-valued mean and correlation function

$$\langle f(t) f(t') \rangle = B_f(t - t').$$

For any individual realization of random force $f(t)$, the solution to Eq. (C.18) has the form

$$x(t) = \int_{t_0}^{t} d\tau f(\tau) e^{-\lambda(t-\tau)}.$$

Consequently, this solution $x(t)$ is also the Gaussian process with the parameters

$$\langle x(t) \rangle = 0, \quad \langle x(t) x(t') \rangle = \int_{t_0}^{t} d\tau_1 \int_{t_0}^{t'} d\tau_2 B_f(\tau_1 - \tau_2) e^{-\lambda(t+t'-\tau_1-\tau_2)}.$$

In addition, we have, for example,

$$\langle f(t) x(t) \rangle = \int_{0}^{t-t_0} d\tau B_f(\tau) e^{-\lambda\tau}.$$

Note that the one-point probability density $P(x, t) = \langle \delta(x(t) - x) \rangle$ of the solution to Eq. (C.18) satisfies the exact equation

$$\left(\frac{\partial}{\partial t} - \lambda \frac{\partial}{\partial x} x \right) P(x, t) = \int_0^{t-t_0} d\tau B_f(\tau) e^{-\lambda \tau} \frac{\partial^2}{\partial x^2} P(x, t), \quad P(x, t_0) = \delta(x),$$

which rigorously follows from Eq. (C.18). As a consequence, we obtain

$$\frac{d}{dt} \langle x^2(t) \rangle = -2\lambda \langle x^2(t) \rangle + 2 \int_0^{t-t_0} d\tau B_f(\tau) e^{-\lambda \tau}.$$

For $t_0 \to -\infty$, process $x(t)$ grades into the stationary Gaussian process with the following one-time statistical parameters ($\langle x(t) \rangle = 0$)

$$\sigma_x^2 = \langle x^2(t) \rangle = \frac{1}{\lambda} \int_0^\infty d\tau B_f(\tau) e^{-\lambda \tau}, \quad \langle f(t) x(t) \rangle = \int_0^\infty d\tau B_f(\tau) e^{-\lambda \tau}.$$

In particular, for exponential correlation function $B_f(t)$,

$$B_f(t) = \sigma_f^2 e^{-|\tau|/\tau_0},$$

we obtain the expressions

$$\langle x(t) \rangle = 0, \quad \langle x^2(t) \rangle = \frac{\sigma_f^2 \tau_0}{\lambda(1 + \lambda \tau_0)}, \quad \langle f(t) x(t) \rangle = \frac{\sigma_f^2 \tau_0}{1 + \lambda \tau_0}, \tag{C.19}$$

which grade into the asymptotic expressions

$$\langle x^2(t) \rangle = \frac{\sigma_f^2 \tau_0}{\lambda}, \quad \langle f(t) x(t) \rangle = \sigma_f^2 \tau_0 \tag{C.20}$$

for $\tau_0 \to 0$.

Multiply now Eq. (C.18) by $x(t)$. Assuming that function $x(t)$ is sufficiently fine function, we obtain the equality

$$x(t) \frac{d}{dt} x(t) = \frac{1}{2} \frac{d}{dt} x^2(t) = -\lambda x^2(t) + f(t) x(t).$$

Averaging this equation over an ensemble of realizations of function $f(t)$, we obtain the equation

$$\frac{1}{2} \frac{d}{dt} \langle x^2(t) \rangle = -\lambda \langle x^2(t) \rangle + \langle f(t) x(t) \rangle, \tag{C.21}$$

whose steady-state solution (it corresponds to the limit process $t_0 \to -\infty$ and $\tau_0 \to 0$)

$$\langle x^2(t) \rangle = \frac{1}{\lambda} \langle f(t)x(t) \rangle$$

coincides with Eqs. (C.19) and (C.20).

Taking into account the fact that $\delta x(t)/\delta f(t-0) = 1$, we obtain the same result for correlation $\langle f(t)x(t) \rangle$ by using the formula

$$\langle f(t)x(t) \rangle = \int\limits_{-\infty}^{t} d\tau B_f(t-\tau) \left\langle \frac{\delta}{\delta f(\tau)} x(t) \right\rangle \tag{C.22}$$

with the effective correlation function

$$B_f^{\mathrm{eff}}(t) = 2\sigma_f^2 \tau_0 \delta(t).$$

Earlier, we mentioned that statistical characteristics of solutions to dynamic problems in the approximation of the delta-correlated random process (field) coincide with the statistical characteristics of the Markovian processes. However, one should clearly understand that this is the case only for statistical averages and equations for these averages. In particular, realizations of process $x(t)$ satisfying the Langevin equation (C.18) drastically differ from realizations of the corresponding Markovian process. The latter satisfies Eq. (C.18) in which function $f(t)$ in the right-hand side is the ideal white noise with the correlation function $B_f(t) = 2\sigma_f^2 \tau_0 \delta(t)$; moreover, this equation must be treated in the sense of generalized functions, because the Markovian processes are not differentiable in the ordinary sense. At the same time, process $x(t)$ — whose statistical characteristics coincide with the characteristics of the Markovian process — behaves as sufficiently fine function and is differentiable in the ordinary sense. For example,

$$x(t)\frac{d}{dt}x(t) = \frac{1}{2}\frac{d}{dt}x^2(t),$$

and we have for $t_0 \to -\infty$ in particular

$$\left\langle x(t)\frac{d}{dt}x(t) \right\rangle = 0. \tag{C.23}$$

On the other hand, in the case of the ideal Markovian process $x(t)$ satisfying (in the sense of generalized functions) the Langevin equation (C.18) with the white noise in the right-hand side, Eq. (C.23) makes no sense at all, and the meaning of the relationship

$$\left\langle x(t)\frac{d}{dt}x(t) \right\rangle = -\lambda \langle x^2(t) \rangle + \langle f(t)x(t) \rangle \tag{C.24}$$

depends on the definition of averages. Indeed, if we will treat Eq. (C.24) as the limit of the equality

$$\left\langle x(t + \Delta) \frac{d}{dt} x(t) \right\rangle = -\lambda \langle x(t) x(t + \Delta) \rangle + \langle f(t) x(t + \Delta) \rangle \tag{C.25}$$

for $\Delta \to 0$, the result will be essentially different depending on whether we use limit processes $\Delta \to +0$, or $\Delta \to -0$. For limit process $\Delta \to +0$, we have

$$\lim_{\Delta \to +0} \langle f(t) x(t + \Delta) \rangle = 2\sigma_f^2 \tau_0,$$

and, taking into account Eq. (C.22), we can rewrite Eq. (C.25) in the form

$$\left\langle x(t + 0) \frac{d}{dt} x(t) \right\rangle = \sigma_f^2 \tau_0. \tag{C.26}$$

On the contrary, for limit process $\Delta \to -0$, we have

$$\langle f(t) x(t - 0) \rangle = 0$$

because of the dynamic causality condition, and Eq. (C.25) assumes the form

$$\left\langle x(t - 0) \frac{d}{dt} x(t) \right\rangle = -\sigma_f^2 \tau_0. \tag{C.27}$$

Comparing Eq. (C.23) with (C.26) and (C.27), we see that, for the ideal Markovian process described by the solution to the Langevin equation with the white noise in the right-hand side and commonly called the *Ohrnstein–Ulenbeck process*, we have

$$\left\langle x(t + 0) \frac{d}{dt} x(t) \right\rangle \neq \left\langle x(t - 0) \frac{d}{dt} x(t) \right\rangle \neq \frac{1}{2} \frac{d}{dt} \langle x^2(t) \rangle.$$

Note that equalities (C.26) and (C.27) can also be obtained from the correlation function

$$\langle x(t) x(t + \tau) \rangle = \frac{\sigma_f^2 \tau_0}{\lambda} e^{-\lambda |\tau|}$$

of process $x(t)$.

To conclude with the discussion of the approximation of the delta-correlated random process (field), we emphasize that, in all further examples, we will treat the sentence like 'dynamic system (equation) with the delta-correlated parameter fluctuations' as the asymptotic limit in which these parameters have temporal correlation radii small in comparison with all characteristic temporal scales of the problem under consideration.

C.3.2 Diffusion Approximation

Applicability of the approximation of the delta-correlated random field $\mathbf{f}(\mathbf{x}, t)$ (i.e., applicability of the Fokker–Planck equation) is restricted by the smallness of the temporal correlation radius τ_0 of random field $\mathbf{f}(\mathbf{x}, t)$ with respect to all temporal scales of the problem under consideration. The effect of the finite-valued temporal correlation radius of random field $\mathbf{f}(\mathbf{x}, t)$ can be considered within the framework of the diffusion approximation. The diffusion approximation appears to be more obvious and physical than the formal mathematical derivation of the approximation of the delta-correlated random field. This approximation also holds for sufficiently weak parameter fluctuations of the stochastic dynamic system and allows describing new physical effects caused by the finite-valued temporal correlation radius of random parameters, rather than only obtaining the applicability range of the delta-correlated approximation. The diffusion approximation assumes that the effect of random actions is insignificant during temporal scales about τ_0, i.e., the system behaves during these time intervals as the free system.

Again, let vector function $\mathbf{x}(t)$ satisfies the dynamic equation (C.1)

$$\frac{d}{dt}\mathbf{x}(t) = \mathbf{v}(\mathbf{x}, t) + \mathbf{f}(\mathbf{x}, t), \quad \mathbf{x}(t_0) = \mathbf{x}_0, \tag{C.28}$$

where $\mathbf{v}(\mathbf{x}, t)$ is the deterministic vector function and $\mathbf{f}(\mathbf{x}, t)$ is the random statistically homogeneous and stationary Gaussian vector field with the statistical characteristics

$$\langle f(\mathbf{x}, t) \rangle = 0, \quad B_{ij}(\mathbf{x}, t; \mathbf{x}', t') = B_{ij}(\mathbf{x} - \mathbf{x}', t - t') = \langle f_i(\mathbf{x}, t) f_j(\mathbf{x}', t') \rangle.$$

Introduce the indicator function

$$\varphi(\mathbf{x}, t) = \delta(\mathbf{x}(t) - \mathbf{x}), \tag{C.29}$$

where $\mathbf{x}(t)$ is the solution to Eq. (C.28) satisfying the Liouville equation (C.6)

$$\left(\frac{\partial}{\partial t} + \frac{\partial}{\partial \mathbf{x}} \mathbf{v}(\mathbf{x}, t) \right) \varphi(\mathbf{x}, t) = -\frac{\partial}{\partial \mathbf{x}} \mathbf{f}(\mathbf{x}, t) \varphi(\mathbf{x}, t). \tag{C.30}$$

As earlier, we obtain the equation for the probability density of the solution to Eq. (C.28)

$$P(\mathbf{x}, t) = \langle \varphi(\mathbf{x}, t) \rangle = \langle \delta(\mathbf{x}(t) - \mathbf{x}) \rangle$$

by averaging Eq. (C.30) over an ensemble of realizations of field $\mathbf{f}(\mathbf{x}, t)$

$$\left(\frac{\partial}{\partial t} + \frac{\partial}{\partial \mathbf{x}} \mathbf{v}(\mathbf{x}, t) \right) P(\mathbf{x}, t) = -\frac{\partial}{\partial \mathbf{x}} \langle \mathbf{f}(\mathbf{x}, t) \varphi(\mathbf{x}, t) \rangle,$$

$$P(\mathbf{x}, t_0) = \delta(\mathbf{x} - \mathbf{x}_0). \tag{C.31}$$

Using the Furutsu–Novikov formula (C.10)

$$\langle f_k(\mathbf{x}, t) R[t; \mathbf{f}(\mathbf{y}, \tau)] \rangle$$

$$= \int d\mathbf{x}' \int dt' B_{kl}(\mathbf{x}, t; \mathbf{x}', t') \left\langle \frac{\delta}{\delta f_l(\mathbf{x}', t')} R[t; \mathbf{f}(\mathbf{y}, \tau)] \right\rangle$$

valid for the correlation between the Gaussian random field $\mathbf{f}(\mathbf{x}, t)$ and arbitrary functional $R[t; \mathbf{f}(\mathbf{y}, \tau)]$ of this field, we can rewrite Eq. (C.31) in the form

$$\left(\frac{\partial}{\partial t} + \frac{\partial}{\partial \mathbf{x}} \mathbf{v}(\mathbf{x}, t) \right) P(\mathbf{x}, t)$$

$$= -\frac{\partial}{\partial x_i} \int d\mathbf{x}' \int_{t_0}^{t} dt' B_{ij}(\mathbf{x}, t; \mathbf{x}', t') \left\langle \frac{\delta}{\delta f_j(\mathbf{x}', t')} \varphi(\mathbf{x}, t) \right\rangle. \qquad (C.32)$$

In the diffusion approximation, Eq. (C.32) is the exact equation, and the variational derivative and indicator function satisfy, within temporal intervals of about temporal correlation radius τ_0 of random field $\mathbf{f}(\mathbf{x}, t)$, the system of dynamic equations

$$\frac{\partial}{\partial t} \frac{\delta \varphi(\mathbf{x}, t)}{\delta f_i(\mathbf{x}', t')} = -\frac{\partial}{\partial \mathbf{x}} \left\{ \mathbf{v}(\mathbf{x}, t) \frac{\delta \varphi(\mathbf{x}, t)}{\delta f_i(\mathbf{x}', t')} \right\},$$

$$\left. \frac{\delta \varphi(\mathbf{x}, t)}{\delta f_i(\mathbf{x}', t')} \right|_{t=t'} = -\frac{\partial}{\partial x_i} \left\{ \delta(\mathbf{x} - \mathbf{x}') \varphi(\mathbf{x}, t') \right\}, \qquad (C.33)$$

$$\frac{\partial}{\partial t} \varphi(\mathbf{x}, t) = -\frac{\partial}{\partial \mathbf{x}} \left\{ \mathbf{v}(\mathbf{x}, t) \varphi(\mathbf{x}, t) \right\}, \quad \varphi(\mathbf{x}, t)|_{t=t'} = \varphi(\mathbf{x}, t').$$

The solution to problem (C.32), (C.33) holds for all times t. In this case, the solution $\mathbf{x}(t)$ to problem (C.28) cannot be considered as the Markovian vector random process because its multi-time probability density cannot be factorized in terms of the transition probability density. However, in asymptotic limit $t \gg \tau_0$, the diffusion-approximation solution to the initial dynamic system (C.28) will be the Markovian random process, and the corresponding conditions of applicability are formulated as smallness of all statistical effects within temporal intervals of about temporal correlation radius τ_0.

C.4 The Simplest Markovian Random Processes

There are only few Fokker–Planck equations that allow an exact solution. First of all, among them are the Fokker–Planck equations corresponding to the stochastic equations that are themselves solvable in the analytic form. Such problems often allow determination of not only the one-point and transitional probability densities, but also the characteristic functional and other statistical characteristics important for practice.

The simplest special case of Eq. (C.11) is the equation that defines the *Wiener random process*. In view of the significant role that such processes plays in physics (for example, they describe the *Brownian motion of particles*), we consider the Wiener process in detail.

C.4.1 Wiener Random Process

The Wiener random process is defined as the solution to the stochastic equation

$$\frac{d}{dt}w(t) = z(t), \quad w(0) = 0,$$

where $z(t)$ is the Gaussian process delta-correlated in time and described by the parameters

$$\langle z(t) \rangle = 0, \quad \langle z(t)z(t') \rangle = 2D\delta(t - t').$$

The solution to this equation

$$w(t) = \int_0^t d\tau z(\tau) \tag{C.34}$$

is the continuous Gaussian nonstationary random process with the parameters

$$\langle w(t) \rangle = 0, \quad \langle w(t)w(t') \rangle = 2D\min(t, t').$$

Figure C.1 shows a realization of the Wiener process (C.34) simulated numerically.

C.4.2 Wiener Random Process with Shear

Consider a more general process that includes additionally the drift dependent on parameter α

$$w(t; \alpha) = -\alpha t + w(t), \quad \alpha > 0.$$

Process $w(t; \alpha)$ is the Markovian process, and its probability density

$$P(w, t; \alpha) = \langle \delta(w(t; \alpha) - w) \rangle$$

Fig. C.1 Realization of the
Wiener process (C.34)

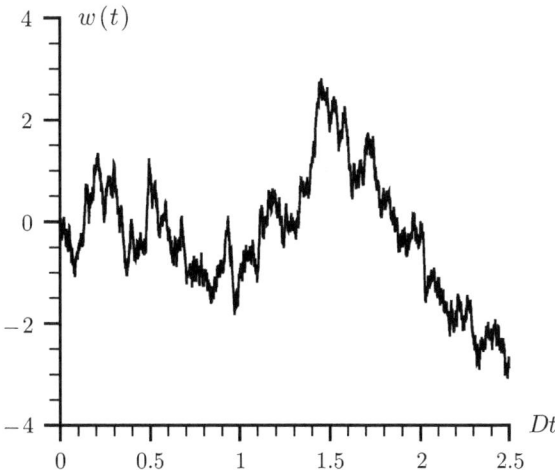

satisfies the Fokker–Planck equation

$$\left(\frac{\partial}{\partial t} - \alpha \frac{\partial}{\partial w}\right) P(w, t; \alpha) = D \frac{\partial^2}{\partial w^2} P(w, t; \alpha), \quad P(w, 0; \alpha) = \delta(w), \quad \text{(C.35)}$$

where $D = \sigma^2 \tau_0$ is the diffusion coefficient. The solution to this equation has the form of the Gaussian distribution

$$P(w, t; \alpha) = \frac{1}{2\sqrt{\pi Dt}} \exp\left\{-\frac{(w + \alpha t)^2}{4Dt}\right\}. \quad \text{(C.36)}$$

The corresponding integral distribution function defined as the probability of the event that $w(t; \alpha) < w$ is given by the formula

$$F(w, t; \alpha) = \int\limits_{-\infty}^{w} dw P(w, t; \alpha) = \text{Pr}\left(\frac{w}{\sqrt{2Dt}} + \alpha\sqrt{\frac{t}{2D}}\right), \quad \text{(C.37)}$$

where function $\text{Pr}(z)$ is the *probability integral* (5.4). In this case, the typical realization curve of the Wiener random process with shear is the linear function of time

$$w^*(t; \alpha) = -\alpha t.$$

In addition to the initial value, supplement Eq. (C.35) with the boundary condition

$$P(w, t; \alpha)|_{w=h} = 0, \quad (t > 0). \quad \text{(C.38)}$$

This condition breaks down realizations of process $w(t; \alpha)$ at the instant they reach boundary h. For $w < h$, the solution to the boundary-value problem (C.35), (C.38) (we denote it as $P(w, t; \alpha, h)$) describes the probability distribution of those realizations of process $w(t; \alpha)$ that survived instant t, i.e., never reached boundary h during the whole temporal interval. Correspondingly, the norm of the probability density appears not unity, but the probability of the event that $t < t^*$, where t^* is the instant at which process $w(t; \alpha)$ reaches boundary h for the first time

$$\int_{-\infty}^{h} dw P(w, t; \alpha, h) = P(t < t^*). \tag{C.39}$$

Introduce the integral distribution function and probability density of random instant at which the process reaches boundary h

$$F(t; \alpha, h) = P(t^* < t) = 1 - P(t < t^*) = 1 - \int_{-\infty}^{h} dw P(w, t; \alpha, h),$$

$$\tag{C.40}$$

$$P(t; \alpha, h) = \frac{\partial}{\partial t} F(t; \alpha, h) = -\frac{\partial}{\partial w} P(w, t; \alpha, h)|_{w=h}.$$

If $\alpha > 0$, process $w(t; \alpha)$ moves on average out of boundary h; as a result, probability $P(t < t^*)$ (C.39) tends for $t \to \infty$ to the probability of the event that process $w(t; \alpha)$ never reaches boundary h. In other words, limit

$$\lim_{t \to \infty} \int_{-\infty}^{h} dw P(w, t; \alpha, h) = P\left(w_{\max}(\alpha) < h\right) \tag{C.41}$$

is equal to the probability of the event that the process absolute maximum

$$w_{\max}(\alpha) = \max_{t \in (0, \infty)} w(t; \alpha)$$

is less than h. Thus, from Eq. (C.41) follows that the integral distribution function of the absolute maximum $w_{\max}(\alpha)$ is given by the formula

$$F(h; \alpha) = P\left(w_{\max}(\alpha) < h\right) = \lim_{t \to \infty} \int_{-\infty}^{h} dw P(w, t; \alpha, h). \tag{C.42}$$

After we solve boundary-value problem (C.35), (C.38) by using, for example, the reflection method, we obtain

$$P(w, t; \alpha, h)$$

$$= \frac{1}{2\sqrt{\pi Dt}} \left\{ \exp\left[-\frac{(w + \alpha t)^2}{4Dt} \right] - \exp\left[-\frac{h\alpha}{D} - \frac{(w - 2h + \alpha t)^2}{4Dt} \right] \right\}. \quad (C.43)$$

Substituting this expression in Eq. (C.40), we obtain the probability density of instant t^* at which process $w(t; \alpha)$ reaches boundary h for the first time

$$P(t; \alpha, h) = \frac{1}{2Dt\sqrt{\pi Dt}} \exp\left\{ -\frac{(h + \alpha t)^2}{4Dt} \right\}.$$

Finally, integrating Eq. (C.43) over w and setting $t \to \infty$, we obtain, in accordance with Eq. (C.42), the integral distribution function of absolute maximum $w_{max}(\alpha)$ of process $w(t; \alpha)$ in the form

$$F(h; \alpha) = P(w_{max}(\alpha) < h) = 1 - \exp\left\{ -\frac{h\alpha}{D} \right\}. \quad (C.44)$$

Consequently, the absolute maximum of the Wiener process has the exponential probability density

$$P(h; \alpha) = \langle \delta(w_{max}(\alpha) - h) \rangle = \frac{\alpha}{D} \exp\left\{ -\frac{h\alpha}{D} \right\}.$$

The Wiener random process offers a possibility of constructing other processes convenient for modeling different physical phenomena. In the case of positive quantities, the simplest approximation of such kind is the logarithmic-normal (lognormal) process. Consider this process in greater detail.

C.4.3 Logarithmic-Normal Random Process

We define the lognormal process (logarithmic-normal process) by the formula

$$y(t; \alpha) = e^{w(t; \alpha)} = \exp\left\{ -\alpha t + \int_0^t d\tau z(\tau) \right\}, \quad (C.45)$$

where $z(t)$ is the Gaussian white noise process with the parameters

$$\langle z(t) \rangle = 0, \quad \langle z(t)z(t') \rangle = 2\sigma^2 \tau_0 \delta(t - t').$$

The lognormal process satisfies the stochastic equation

$$\frac{d}{dt} y(t; \alpha) = \{-\alpha + z(t)\} y(t; \alpha), \quad y(0; \alpha) = 1.$$

The one-time probability density of the lognormal process is given by the formula

$$P(y, t; \alpha) = \langle \delta (y(t; \alpha) - y) \rangle = \langle \delta \left(e^{w(t;\alpha)} - y \right) \rangle$$

$$= \frac{1}{y} \langle \delta (w(t; \alpha) - \ln y) \rangle = \frac{1}{y} P(w, t; \alpha)|_{w=\ln y},$$

where $P(w, t; \alpha)$ is the one-time probability density of the Wiener process with a drift, which is given by Eq. (C.36), so that

$$P(y, t; \alpha) = \frac{1}{2y\sqrt{\pi Dt}} \exp \left\{ -\frac{(\ln y + \alpha t)^2}{4Dt} \right\}$$

$$= \frac{1}{2y\sqrt{\pi Dt}} \exp \left\{ -\frac{\ln^2 \left(y e^{\alpha t} \right)}{4Dt} \right\}, \qquad (C.46)$$

where $D = \sigma^2 \tau_0$.

Note that the one-time probability density of random process $\widetilde{y}(t; \alpha) = 1/y(t; \alpha)$ is also lognormal and is given by the formula

$$P(\widetilde{y}, t; \alpha) = \frac{1}{2\widetilde{y}\sqrt{\pi Dt}} \exp \left\{ -\frac{\ln^2 \left(\widetilde{y} e^{-\alpha t} \right)}{4Dt} \right\}, \qquad (C.47)$$

which coincides with Eq. (C.46) with parameter α of opposite sign. Correspondingly, the integral distribution functions are given, in accordance with Eq. (C.37), by the expressions

$$F(y, t; \alpha) = P(y(t; \alpha) < y) = \Pr \left(\frac{1}{\sqrt{2Dt}} \ln \left(y e^{\pm \alpha t} \right) \right), \qquad (C.48)$$

where $\Pr(z)$ is the probability integral (5.4)

Figure 5.1 show the curves of the lognormal probability densities for $\alpha/D = 1$ and dimensionless times $\tau = Dt = 0.1$ and 1. Figure 5.2 shows these probability densities at $\tau = 1$ in logarithmic scale along the abscissa.

Structurally, these probability distributions are absolutely different. The only common feature of these distributions consists in the existence of long flat *tails* that appear in distributions at $\tau = 1$; these tails increase the role of high peaks of processes $y(t; \alpha)$ and $\widetilde{y}(t; \alpha)$ in the formation of the one-time statistics.

Having only the one-point statistical characteristics of processes $y(t; \alpha)$ and $\widetilde{y}(t; \alpha)$, one can obtain a deeper insight into the behavior of realizations of these processes on the whole interval of times $(0, \infty)$. In particular,

(1) The lognormal process $y(t; \alpha)$ is the Markovian process and its one-time probability density satisfies the Fokker–Planck equation

$$\left(\frac{\partial}{\partial t} - \alpha\frac{\partial}{\partial y}y\right)P(y, t; \alpha) = D\frac{\partial}{\partial y}y\frac{\partial}{\partial y}yP(y, t; \alpha), \quad P(y, 0; \alpha) = \delta(y - 1). \quad \text{(C.49)}$$

From Eq. (C.49), one can easily derive the equations for the moment functions of processes $y(t; \alpha)$ and $\widetilde{y}(t; \alpha)$; solutions to these equations are given by the formulas $(n = 1, 2, \ldots)$

$$\langle y^n(t; \alpha)\rangle = e^{n(n-\alpha/D)Dt}, \quad \langle \widetilde{y}^n(t; \alpha)\rangle = \left\langle\frac{1}{y^n(t; \alpha)}\right\rangle = e^{n(n+\alpha/D)Dt}, \quad \text{(C.50)}$$

from which follows that moments exponentially grow with time.

From Eq. (C.49), one can easily obtain the equality

$$\langle \ln y(t)\rangle = -\alpha t.$$

Consequently, parameter α can be rewritten in the form

$$-\alpha = \frac{1}{t}\langle \ln y(t)\rangle \quad \text{or} \quad \alpha = \frac{1}{t}\langle \ln \widetilde{y}(t)\rangle. \quad \text{(C.51)}$$

Note that many investigators give great attention to the approach based on the Lyapunov analysis of stability of solutions to deterministic ordinary differential equations

$$\frac{d}{dt}\mathbf{x}(t) = A(t)\mathbf{x}(t).$$

This approach deals with the upper limit of problem solution

$$\lambda_{\mathbf{x}(t)} = \overline{\lim_{t\to+\infty}}\,\frac{1}{t}\ln|\mathbf{x}(t)|$$

called the characteristic index of the solution. In the context of this approach applied to stochastic dynamic systems, these investigators often use statistical analysis at the last stage to interpret and simplify the obtained results; in particular, they calculate statistical averages such as

$$\langle\lambda_{\mathbf{x}(t)}\rangle = \overline{\lim_{t\to+\infty}}\,\frac{1}{t}\langle \ln|\mathbf{x}(t)|\rangle. \quad \text{(C.52)}$$

Parameter α is the *Lyapunov exponent* of the lognormal random process $y(t)$.

(**2**) From the integral distribution functions, one can calculate the typical realization curves of lognormal processes $y(t; \alpha)$ and $\widetilde{y}(t; \alpha)$

$$y^*(t) = e^{\langle \ln y(t)\rangle} = e^{-\alpha t}, \quad \widetilde{y}^*(t) = e^{\langle \ln \widetilde{y}(t)\rangle} = e^{\alpha t}, \quad \text{(C.53)}$$

which are the exponentially decaying curve in the case of process $y(t; \alpha)$ and the exponentially increasing curve in the case of process $\widetilde{y}(t; \alpha)$.

Consequently, the exponential increase of moments of random processes $y(t; \alpha)$ and $\widetilde{y}(t; \alpha)$ are caused by deviations of these processes from the typical realization curves $y^*(t; \alpha)$ and $\widetilde{y}^*(t; \alpha)$ towards both large and small values of y and $\widetilde{y}$.

As it follows from Eq. (C.50) at $\alpha/D = 1$, the average value of process $y(t; D)$ is independent of time and is equal to unity. Despite this fact, according to Eq. (C.48), the probability of the event that $y < 1$ for $Dt \gg 1$ rapidly approaches the unity by the law

$$P(y(t; D) < 1) = \Pr\left(\sqrt{\frac{Dt}{2}}\right) = 1 - \frac{1}{\sqrt{\pi Dt}} e^{-Dt/4},$$

i.e., the curves of process realizations run mainly below the level of the process average $\langle y(t; D)\rangle = 1$, which means that namely large peaks of the process govern the behavior of statistical moments of process $y(t; D)$. n Here, we have a clear contradiction between the behavior of statistical characteristics of process $y(t; \alpha)$ and the behavior of process realizations.

(3) The behavior of realizations of process $y(t; \alpha)$ on the whole temporal interval can also be evaluated with the use of the p-majorant curves $M_p(t, \alpha)$. We call the majorant curve the curve $M_p(t, \alpha)$ for which inequality $y(t; \alpha) < M_p(t, \alpha)$ is satisfied for all times t with probability p, i.e.,

$$\mathsf{P}\left\{y(t; \alpha) < M_p(t, \alpha) \text{ for all } t \in (0, \infty)\right\} = p.$$

The above statistics (C.44) of the absolute maximum of the Wiener process with a drift $w(t; \alpha)$ makes it possible to outline a wide enough class of the majorant curves. Indeed, let p be the probability of the event that the absolute maximum $w_{\max}(\beta)$ of the auxiliary process $w(t; \beta)$ with arbitrary parameter β in the interval $0 < \beta < \alpha$ satisfies inequality $w(t; \beta) < h = \ln A$. It is clear that the whole realization of process $y(t; \alpha)$ will run in this case below the majorant curve

$$M_p(t, \alpha, \beta) = A e^{(\beta - \alpha)t} \tag{C.54}$$

with the same probability p. As may be seen from Eq. (C.44), the probability of the event that process $y(t; \alpha)$ never exceeds majorant curve (C.54) depends on this curve parameters according to the formula

$$p = 1 - A^{-\beta/D}.$$

This means that we derived the one-parameter class of exponentially decaying majorant curves

$$M_p(t, \alpha, \beta) = \frac{1}{(1 - p)^{D/\beta}} e^{(\beta - \alpha)t}. \tag{C.55}$$

Fig. C.2 Schematic behaviors of a realization of process $y(t; D)$ and majorant curve $M(\tau)$ (C.56)

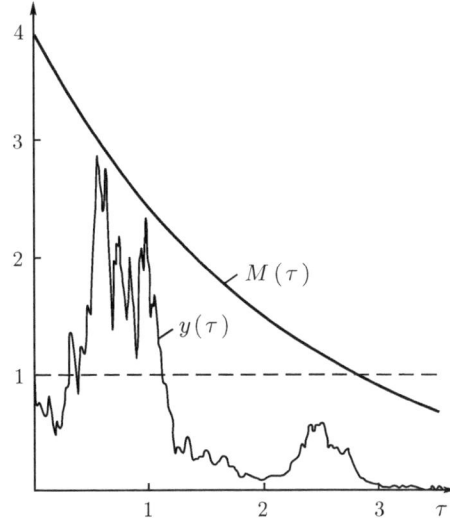

Notice the remarkable fact that, despite statistical average $\langle y(t; D) \rangle = 1$ remains constant and higher-order moments of process $y(t; D)$ are exponentially increasing functions, one can always select an exponentially decreasing majorant curve (C.55) such that realizations of process $y(t; D)$ will run below it with arbitrary predetermined probability $p < 1$. In particular, inequality $(\tau = Dt)$

$$y(t; D) < M_{1/2}(t, D, D/2) = M(\tau) = 4e^{-\tau/2} \tag{C.56}$$

is satisfied with probability $p = 1/2$ for any instant t from interval $(0, \infty)$.

Figure C.2 schematically shows the behaviors of a realization of process $y(t; D)$ and the majorant curve (C.56). This schematic is an additional fact in favor of our conclusion that the exponential growth of moments of process $y(t; D)$ with time is the purely statistical effect caused by averaging over the whole ensemble of realizations.

Note that the area below the exponentially decaying majorant curves has a finite value. Consequently, high peaks of process $y(t; \alpha)$, which are the reason of the exponential growth of higher moments, only insignificantly contribute to the area below realizations; this area appears finite for almost all realizations, which means that the peaks of the lognormal[1] process $y(t; \alpha)$ are sufficiently narrow.

---

[1] Sentence by S.I. Vavilov from his paper [2, p. 584].

Bibliography

1. Shcherbakov, R.N.: 'Galileo was amazingly endowed with a gift of.. inculcating scientific truth' (Sentence by S.I. Vavilov from his paper [2]). Phys. Usp. **57**(2), 143–151 (2014)
2. Vavilov, S.I.: Galileo in history of optics. UFN **83**(8), 583–615 (1964) (in Russian)
3. Nicolis, G., Prigogin, I.: Exploring Complexity: An Introduction. W.H. Freeman and Co., New York (1989)
4. Klyatskin, V.I.: On Statistical theory of two-dimensional turbulence. J. Appl. Math. Mech. **33**(5), 864–866 (1969)
5. Klyatskin, V.I.: Equilibrium states for quasigeostrophic flows with random topography. Izv. Atmos. Ocean. Phys. **31**(6), 717–722 (1995)
6. Klyatskin, V.I., Gurarie, D.: Equilibrium states for quasigeostrophic flows with random topography. Phys. D **98**, 466–480 (1996)
7. Klyatskin, V.I.: Equilibrium distributions for hydrodynamic flows. Discontin. Nonlinearity Complex. **4**(3), 243–256 (2015)
8. Landau, L.D., Lifshitz, E.M.: Fluid Mechanics, 2nd edn. Pergamon Press, London (1987)
9. Hopf, E., Titt, E.W.: On certain special solution of the equation of statistical hydrodynamics. J. Ration. Mech. Anal. **2**, 587 (1953)
10. Kraichnan, R.H., Montgomery, D.: Two-dimensional turbulence. Rep. Prog. Phys. **43**, 547–619 (1980)
11. Hopfinger, E.J., Browand, F.K.: Vortex solitary waves in rotating, turbulent flow. Nature **295**(5848), 393–394 (1982)
12. Griffiths, R.W., Hopfinger, E.J.: The structure of mesoscale turbulence and horizontal spreading at ocean fronts. Deep Sea Res. A **31**(3), 245–269 (1984)
13. Boubnov, B.M., Golitsyn, G.S.: Experimental study of convective structures in rotating fluids. J. Fluid Mech. **167**(6), 503–531 (1986)
14. Hopfinger, E.J.: Turbulence and vortices in rotating fluids. In: Germain, P., Piau, M., Caillerie, D. (eds.) Theoretical and Applied Mechanics. IUTAM, pp. 117–138. Elsevier Science Publishers B.V, North-Holland (1989)
15. Boubnov, B.M., Golitsyn, G.S.: Convection in rotating fluids. Fluid Mechanics and its Applications, vol. 29. Kluver Academic Publishers, Dordrecht (1995)
16. Golitsyn, G.S.: Statistics and Dynamics of Processes and Phenomena in Nature: Methods, Tools, and Results. KRASAND, Moscow (2013) (in Russian)
17. Pavlov, V., Buisine, D., Goncharov, V.: Formation of vortex clusters on a sphere. Nonlinear Proc. Geophys. **8**(8), 9–19 (2001)

© Springer International Publishing AG 2017

V.I. Klyatskin, *Fundamentals of Stochastic Nature Sciences*,
Understanding Complex Systems, DOI 10.1007/978-3-319-56922-2

18. Karimova, S.S., Lavrova, OYu., Solov'ev, D.M.: Observation of Eddy structures in the Baltic sea with the use of radiolocation and radiometric sattelite data. Izvest. Atmos. Ocean. Phys. **49**(9), 1006–1013 (2011)

19. Karimova, S.: Spiral eddies in the Baltic, Black and Caspian seas as seen by satellite radar data. Adv. Space Res. **50**, 1107–1124 (2012)

20. Obukhov, A.M.: The Kolmogorov flow and its laboratory modeling. Uspekhi Mat. Nauk **38**(4(232)), 101–111 (1983)

21. Obukhov, A.M.: Turbulence and Dynamics of Atmosphere. Gidrometizdat, Leningrad (1988) (in Russian)

22. Meshalkin, L.D., Sinai, Ya.G.: Investigation of the stability of a stationary solution of a system of equations for the plane movement of an incompressible viscous flow. J. Appl. Math. Mech. **25**(6), 1700–1705 (1961)

23. Yudovich, V.I.: Example of the birth of secondary steady-state or periodic flow at stability failure of the laminar flow of viscous noncompressible fluid. J. Appl. Math. Mech. **29**(3), 527–544 (1965)

24. Dolzhansky, F.V., Klyatskin, V.I., Obukhov, A.M., Chusov, M.A.: Nonlinear Hydrodynamic Type Systems. Nauka, Moscow (1974) (in Russian)

25. Gledzer, E.B., Dolzhanskii, F.V., Obukhov, A.M.: Systems of Hydrodynamical Type and Their Applications. Nauka, Moscow (1981) (in Russian)

26. Klyatskin, V.I.: To the nonlinear theory of stability of periodic flow. J. Appl. Math. Mech. **36**(2) (1972)

27. Dolzhansky, F.V.: Fundamentals of geophysical hydrodynamics. Encyclopedia of Mathematical Sciences. Mathematical Physics IV, vol. 103. Springer, Berlin (2013)

28. Klyatskin, V.I.: Lectures on Dynamics of Stochastic Systems. Elsevier, Boston (2010)

29. Klyatskin, V.I.: Stochastic Equations: Theory and Applications in Acoustics, Hydrodynamics, Magnetohydrodynamics, and Radiophysics, vols. 1, 2. Springer, Helderberg (2015)

30. Klyatskin, V.I.: Spatial structures can form in stochastic dynamic systems due to near-zero-probability events: (comment on '21st century: what is life from the perspective of physics?'). Phys. Usp. **55**(11), 1152–1154 (2012)

31. Klyatskin, V.I.: On the criterion of stochastic structure formation in random media. In: Proceedings of the 4th International Interdisciplinary Chaos Symposium, pp. 69–73. Springer, Berlin (2013)

32. Klyatskin, V.I.: Clustering of random positive field as a law of nature. Teor. Math. Phys. **176**(3), 1252–1256 (2013)

33. Klyatskin, V.I.: On the statistical theory of spatial structure formation in random media. Russ. J. Math. Phys. **20**(3), 295–314 (2013)

34. Klyatskin, V.I.: Clustering of a positive random field - what is it? Discontin. Nonlinearity Complex. **4**(3), 235–242 (2015)

35. Klyatskin, V.I.: Stochastic structure formation in random media. Phys.-Usp. **59**(1), 67–95 (2016)

36. Klyatskin, V.I.: Integral characteristics: a key to understanding structure formation in stochastic dynamical systems. Phys.-Usp. **54**(5), 441–464 (2011)

37. Zel'dovich, Ya.B., Molchanov, S.A., Ruzmaikin, A.A., Sokolov, D.D.: Intermittency of passive fields in random media. Sov. Phys. JETP **62**, 1188–1194 (1985)

38. Zel'dovich, Ya.B., Molchanov, S.A., Ruzmaikin, A.A., Sokolov, D.D.: Intermittency in random media. Sov. Phys. Usp. **30**, 353–369 (1987)

39. Batchelor, G.K., Townsend, A.A.: Small-scale variation of convected quantities like temperature in turbulent fluid. Proc. R. Soc. Lond. Ser. A. **30**(2), 238–255 (1949)

40. Batchelor, G.K., Howells, I.D., Townsend, A.A.: Small-scale variation of convected quantities like temperature in turbulent fluid. J. Fluid Mech. **5**(1), 113–139 (1959)

41. Monin, A.S., Yaglom, A.M.: Statistical Fluid Mechanics. MIT Press, Cambridge (1980)

42. Zakharov, V.E.,: 'Crazy waves'. http://www.youtube.com/watch?v=Bt3ZOswGd-4; 'Roque Waves'. http://www.zoomby.ru/watch/114832-academia (in Russian)

43. Zakharov, V.E., Shamin, R.V.: Statistics of rogue waves in computer experiments. JETP Lett. **96**(1), 66–69 (2012)
44. Shamin, R.V.: Mathematical issues of rogue waves. LENAND, Moscow (2016) (in Russian)
45. Ruzmaikin, A.: Climate as a game of chance. Phys.-Usp. **57**(3), 289–294 (2014)
46. Anderson, P.W.: Absence of diffusion in certain random lattices. Phys. Rev. **109**, 1492–1505 (1958)
47. Lifshits, I.M., Gredeskul, S.A., Pastur, L.A.: Introduction to the Theory of Disordered Solids. Wiley, New York (1988)
48. Klyatskin, V.I., Saichev, A.I.: Statistical and dynamical localization of plane waves in randomly layered media. Sov. Phys. Usp. **35**(3), 231–247 (1992)
49. Klyatskin, V.I., Saichev, A.I.: Statistical theory of the diffusion of a passive tracer in a random velocity field. JETP **84**(4), 716–724 (1997)
50. Mikhailov, A.S., Uporov, I.V.: Critical phenomena in media with breeding, decay, and diffusion. Sov. Phys. Usp. **27**(9), 695–714 (1984)
51. Klyatskin, V.I., Chkhetiani, O.G.: On the diffusion and clustering of a magnetic field in random velocity fields. JETP **109**(2), 345–356 (2009)
52. Landau, L.D., Lifshitz, E.M.: Course of theoretical physics. Electrodynamics of Continuous Media, vol. 8. Butterworth-Heinemann, Oxford (1984)
53. Malyanov, D.: http://www.gazeta.ru/science/2011/06/17_a_3664677.shtml; Atkinson N.: http://www.universetoday.com/86446/voyagers-find-giant-jacuzzi-like-bubbles-at-edge-of-solar-system/
54. Klyatskin, V.I.: Propagation of electromagnetic waves in randomly inhomogeneous media as a problem of statistical mathematical physics. Phys. Usp. **47**(2), 169–186 (2004)
55. Klyatskin, V.I.: Statistical topography and Lyapunov's exponents in dynamic stochastic systems. Phys. Usp. **51**(4), 395–407 (2008)
56. Klyatskin, V.I.: Modern methods for the statistical description of dynamical stochastic sestems. Phys. Usp. **52**(5), 514–519 (2009)
57. Klyatskin, V.I., Koshel', K.V.: The field of a point source in a layered media, Dokl. Acad. Nauk SSSR, **288**(6), 1478–1481 (1986); Transactions (doclady) USSR Acad. Sciences, Earth Science Section (1986)
58. Klyatskin, V.I., Yaroshchuk, I.O.: Fluctuations of intensity of a wave in randomly inhomogeneous edium. VII. Numerical modeling wave propagation in a stochastic media. Radiophys. Quantum Electron. **26**(10), 900–907 (1983)
59. Zavorotnyi, V.U., Klyatskin, V.I., Tatarskii, V.I.: Strong fluctuations of the intensity of electromagnetic waves in randomly inhomogeneous media. Sov. Phys. JETP **46**(2), 252–260 (1977)
60. Klyatskin, V.I.: Ondes et Équations Stochastiques dans les milieus Aléatoirement non Homogènes. Les éditions de Physique, Besançon-Cedex (1985) (in French)
61. Klyatskin, V.I., Yakushkin, I.G.: Statistical theory of the propagation of optical radiation in turbulent media. JETP **84**(6), 1114–1121 (1997)
62. Klyatskin, V.I., Elperin, T.: Clustering of the low-inertia particle number density field in random divergence-free hydrodynamic flows. JETP **95**(2), 328–340 (2002)
63. Maxey, M.R.: The gravitational settling of aerosol particles in homogeneous turbulence and random flow field. J. Fluid Mech. **174**, 441–465 (1987)
64. Klyatskin, V.I.: Diffusion and clustering of sedimenting tracer in random hydrodynamic flows. JETP **99**(5), 1005–1017 (2004)
65. Izrael', Yu.A.: Radioactive Contamination of the Natural Media as the Result of the Incident at the Chernobyl Nuclear Power Plant. Komtekhprint, Moscow (2006). http://www.ibrae.ac.ru/images/stories/ibrae/chernobyl/israel.pdf
66. Gurbatov, S.N., Saichev, A.I., Shandarin, S.F.: Large-scale structure of the Universe. The Zel'dovich approximation and the adhesion model. Phys.– Usp. **55**(3), 223–249 (2012)
67. Klyatskin, V.I., Koshel', K.V.: Impact of diffusion on surface clustering in random hydrodynamic flows. Phys. Rev. E **95**(013109), 1–7 (2017)
68. Rytov, S.M., Kravtsov, Yu.A., Tatarskii, V.I.: Principles of Statistical Radiophysics, vols. 1–4. Springer, Berlin (1987–1989)

69. Klyatskin, V.I.: Stochastic dynamo in critical situation. Theor. Math. Phys. **172**(3), 1243–1262 (2012)
70. Klyatskin, V.I., Yakushkin, I.G.: Stochastic transport in random wave fields. JETP **91**(4), 736–747 (2000)
71. Klyatskin, V.I., Koshel', K.V.: The simplest example of the development of a cluster-structured passive tracer field in random flows. Phys. Usp. **43**(7), 717–723 (2000)
72. Klyatskin, V.I.: Integral characteristics: a key to understanding structure formation in stochastic dynamical systems. Phys. Usp. **54**(5), 441–464 (2011)
73. Kharif, C., Pelinovskyy, E., Slyunaen, A.: Rogue Waves in the Ocean. Springer, Berlin (2009)
74. Dysthe, K., Krogstad, H.E., Muller, P.: Oceanic rogue waves. Annu. Rev. Fluid Mech. **40**, 287–310 (2008)
75. Christou, M., Ewans, K.: Field Measurements of rogue water waves. J. Phys. Oceanogr. **44**(9), 2317–2355 (2014). doi:10.1175/JPO-D-13-0199.1]
76. Ruban, V.P.: Enhanced rise of rogue waves in alant wave groups. Pis'ma v ZhETF **94**(3), 177–181 (2011)
77. Ruban, V.P.: On the nonlinear Schrödinger equation for waves on a nonuniform current. Pis'ma v ZhETF **95**(9), 486–491 (2012)
78. Ruban, V.P.: On the modulation instability of surface waves on a large-scale shear flow. Pis'ma v ZhETF **97**(4), 188–193 (2013)
79. Ruban, V.P.: Rogue waves at low Benjamin-Feir indeces: numerical study of the role of non-linearity. Pis'ma v ZhETF **97**(12), 686–690 (2013)
80. http://en.wikipedia.org/wiki/File:Ile_de_re.JPG
81. Metzger, J.J., Fleischmann, R., Geisel, T.: Statistics of extreme waves in random media. Phys. Rev. Lett. **11**(5), 203903 (2014)
82. Klyatskin, V.I.: Anomalous waves as an object of statistical topography: problem statement. Teor. i Mat. Fiz. **180**(1), 850–861 (2014)
83. Klyatskin, V.I., Koshel', K.V.: Anomalous sea surface structures as an object of statistical topography. Phys. Rev. E **91**(063003), 1–12 (2015)
84. Klyatskin, V.I., Koshel', K.V.: Statistical structuring theory in parametrically excitable dynamical systems with a Gaussian pump. Teor. i Mat. Fiz. **186**(3), 411–429 (2016)